THIS BOOK COULD
SAVE YOUR LIFE

Graham Lawton has a degree in biochemistry and an MSc in science communication, both from Imperial College, London. He is an award-winning staff writer at *New Scientist*. He lives in London.

Books by *New Scientist* include

How to be Human
The Origin of (almost) Everything
The Brain
How Long is Now?
The Universe Next Door
This Book Will Blow Your Mind

THIS
BOOK
COULD
SAVE
YOUR
LIFE

The Science of Living
Longer Better

GRAHAM LAWTON

New
Scientist

First published in Great Britain in 2020 by John Murray (Publishers)
First published in the United States of America in 2020
by Nicholas Brealey Publishing
Imprints of John Murray Press
An Hachette UK company

This paperback edition published in 2021

1

Copyright © New Scientist 2020

This book is intended for information purposes only, and should not
be taken as individual medical advice. If you have medical concerns,
you should consult a medical practitioner. You should also consult
a medical practitioner before making changes to your diet or exercise
regime, especially if you have pre-existing health conditions.

A CIP catalogue record for this title is available from the British Library

Paperback ISBN 978-1-529-31131-0
UK eBook ISBN 978-1-529-31132-7
US eBook ISBN 978-1-529-36208-4

Typeset in Bembo by Palimpsest Book Production Ltd, Falkirk, Stirlingshire

Printed and bound in Great Britain by Clays Ltd, Elcograf S.p.A.

John Murray policy is to use papers that are natural, renewable and
recyclable products and made from wood grown in sustainable forests.
The logging and manufacturing processes are expected to conform
to the environmental regulations of the country of origin.

John Murray (Publishers)
Carmelite House
50 Victoria Embankment
London EC4Y 0DZ
www.johnmurraypress.co.uk

Nicholas Brealey Publishing US
Hachette Book Group
Market Place Center, 53 State Street,
Boston, MA 02109 USA
www.nbuspublishing.com

CONTENTS

INTRODUCTION

I HAVE A confession to make: I'm an overweight, lazy slob. I love beer and fast food and the odd crafty fag outside a pub. I can't remember the last time I drank fewer than fourteen units of alcohol in a week. I have a gym membership but I rarely use it. After work I slump in front of the TV, with snacks. Sometimes I drive to a shop I could easily walk to, usually to buy beer. I take medication to keep my blood pressure under control and my body mass index (BMI) puts me in the 'over-weight' category. Some lifestyle guru!

But here's another thing. Last year I cycled over 1,200 miles and ran the equivalent of fifteen marathons. I eat my five portions of fruit and vegetables a day, stay hydrated, watch my salt intake and don't eat meat. I fast regularly, go outdoor swimming, do Pilates and walk up eight flights of stairs to my desk instead of taking the lift. And despite my bad habits, my good ones seem to be winning. I'm fifty, but my 'biological age' was recently measured as forty-five. I'm living proof that you don't have to be a health and fitness fanatic to be reason-ably healthy and fit.

What's my secret? I'm a science journalist with a background in biochemistry, and I have spent much of my working life keeping abreast of the latest thinking in biomedical science and writing about it for a non-specialist audience. As a result I'm able to evaluate health claims about nutrition, exercise, dietary supplements and more. I can spot a sensational headline a mile off, sniff out a fad and separate fact from fiction.

And I put what I know into practice. I'm not claiming to have invented the ideal health and fitness regime, but I lead a healthy but not too restrictive life, based on evidence. And that is the purpose of this book: to empower you to do the same. And I really mean 'empower'. Good health is something we should all strive for, because the opposite – ill health – is not pleasant, and will eventually kill you.

Knowing what to do isn't easy. Biomedical science advances all the time, which is why we now routinely live to grand old ages and enjoy better health than our grandparents. But with this increase in knowledge come greater complexity and the risk of information overload. There's more advice out there than ever competing for your attention, but not always from the most authoritative sources.

My aim is to cut through the noise. I have rounded up the latest and most rigorous health research and converted it into useful, actionable advice on all the big health questions: we'll cover nutrition, diet, weight loss, hydration, exercise, preventative medicine, sleep and ageing. On the way I'll explain the basic science, debunk common myths, give you the tools you need to evaluate critically claims and counterclaims and help you to see beyond the headlines. Think of it as a manual for a healthy lifestyle.

If you have specific goals such as losing weight, improving your diet, getting fit, sleeping better or knowing which dietary supplements or superfoods are worth the money, you'll find advice in here. But the book is more than the sum of its parts. A mantra in health science is that we must consider the 'totality of the evidence'. That is what you'll get – a comprehensive and evidence-based guide to a heathier, fitter and, I hope, longer life.

Of course, lots of other people make similar claims. The world is not short of advice about what to eat and drink, how to stay fit, how to sleep better and live longer. But most of it is based on little more than myths, wishful thinking or unscientific mumbo-jumbo.

You can take this advice if you like. Or you can put your trust in the people whose job it is to work out what actually works: scientists. For every celebrity with a plan there are hundreds of nutritionists, exercise physiologists, sleep scientists and biomedical researchers doing the research and translating it into action. You won't read about it in glossy magazines or wellness blogs or see it on TV – it is mostly found in the pages of serious and technical scientific journals – but if you're really interested in living a longer, healthier and happier life, this is the stuff you need to know.

First, though, a disclaimer. Contrary to what many self-appointed gurus will tell you, the road to health, fitness and wellness is neither simple nor easy. The science is often uncertain, contradictory or difficult to translate into concrete advice. Changing your lifestyle for the better inevitably requires some effort, willpower and sacrifice. If you want quick and easy fixes, this book is not for you. (But before you put it down and go in search of a quick and easy fix, let me tell you – for free – that anyone promising one is taking you for a fool and trying to part you from your money.) Do not let the certainty of sacrifice put you off. A modicum of effort can pay big dividends. And even just knowing that advice is based on good science can be a strong motivator to act.

That is also what sets this book apart. Resolving to change your life for the better is easy; actually changing it is hard. We've all made a resolution in January only to fall off the treadmill come February. But again, science can help. Understanding willpower, motivation and habit formation and how to harness them to your benefit is a big part of the battle.

And I promise it will pay off. You only get one life, and it's shrinking every day. If you enjoy being alive and well and want more of it, this is the book for you.

Sorting fact from fiction

Before we tuck into our main course of health advice, it is a good idea to have a starter of statistics: the science of drawing valid conclusions from raw data. Understanding a bit of mathematics can be key to debunking a sensationalist headline.

Consider the claim that taking multivitamin supplements prevents cancer. To see whether this holds up, you need to run an experiment where one group of people takes multivitamins and another doesn't. After a certain amount of time, you see how many cases of cancer occurred in the two groups. This is your raw data. To turn it into a scientifically valid conclusion, you need to put it through the statistical wringer.

You might think that if there were more cases of cancer in the non-vitamin group you've proved your point. But you haven't – the result could be due to chance. To rule this out you need to know the 'statistical significance' of the difference between the two groups. There is a standard equation for calculating this, which we don't need to concern ourselves with here. Suffice to say that the answer will be a number between 0 and 1.

For a result to be considered significant, that number needs to be at least 0.95. That means there is a 95 per cent probability that it did not happen by chance, and therefore reflects reality.

Some studies impose a higher threshold of 0.99 but 0.95 is the minimum standard of proof to look for. Bear in mind what that actually means: there is a one in twenty chance that it is a fluke. This is why multiple successful trials are needed to convince medical authorities that an effect is real.

Even if a finding is 99 per cent certain, that means there is a 1 per cent chance that it isn't. But that is not a licence to dismiss all scientific findings as 99 per cent certain means what it says: it is overwhelmingly likely to be true.

Statistical significance isn't everything. The second tip is to look at sample sizes – the number of people in your study. The bigger a sample size the more likely the result is to be valid. It's a bit like tossing a coin. Five tosses might give you four heads and a tail, but you'd be a fool to conclude that the probability of getting a head is therefore 0.8. Keep tossing the coin, say 100 times, and (assuming the coin is fair) that initial statistical noise will cancel out and a probability much closer to 0.5 will emerge.

Large sample sizes are also required to reveal small differences between intervening and doing nothing. If a nutritional study has a very low number of participants, say twenty or even fifty, take any conclusion with a very large pinch of salt.

The most important number to emerge from statistical analysis, however, is risk. In our example, that would be the risk of not taking multivitamins versus the risk of taking them.

But let's use a different example. Say you come across the finding that women who use talcum powder are 40 per cent more likely to develop ovarian cancer. Scary or what? It is easy to interpret (or spin) this discovery as meaning that anyone using talc daily has a 40 per cent chance of ovarian cancer.

It doesn't, because 40 per cent is a 'relative risk' – the extra risk that you are taking by using talc. It means little unless you know the absolute risk, or how likely you are to get ovarian cancer if you don't use talc. That number is about 27 per 100,000, or 0.027 per cent. So a 40 per cent increase raises that to 38 per 100,000 – not negligible, but nothing to panic about.

Such rules of thumb can help anywhere you see a statistical claim. They have their limits – they are powerless to reveal when someone has cherry-picked their data or massaged their figures. But they are a good start when sorting out damned lies from statistics.

This book is intended for information purposes only and should not be taken as individual medical advice. If you have medical concerns you should consult a medical practitioner. You should also consult a medical practitioner before making changes to your diet or exercise regime, especially if you have pre-existing health conditions.

THE TRUTH ABOUT FOOD

THERE'S A FAMOUS scene in Woody Allen's 1973 film *Sleeper* in which two scientists in the year 2173 are discussing the dietary advice of the late twentieth century. 'You mean there was no deep fat, no steak, or cream pies, or hot fudge?' asks one, incredulous. 'Those were thought to be unhealthy,' replies the other. 'Precisely the opposite of what we now know to be true.'

'Incredible,' says the first.

We're not quite in incredible territory yet, but deep fat and cream pies are starting to look a lot less unhealthy than they once did. In the past few years, saturated fat – once the pariah of your plate – has been subject to a serious rethink.

And it is not just fat. The early twenty-first century has been a period of upheaval and soul searching for nutrition science. Almost everything we thought we knew has been challenged, and some of it overturned. Food groups once considered unhealthy are being rehabilitated, and vice versa.

This is probably familiar to anyone who keeps an eye on the latest news about diet and health with the goal of trying to eat well. It's confusing. For some reason the advice seems to keep on changing, so you end up not knowing if you are eating the right things.

One thing, however, remains absolutely beyond doubt. You are what you eat. Diet has a huge influence on your heath and is the bedrock of a healthy lifestyle. This chapter will bring you up to date on the latest thinking about some of the major food groups and nutrients, from fats and sugar to salt, meat, dairy and gluten, and conclude with some take-home messages.

But first, a health warning. Nutrition science is hard to do well, and rarely produces definitive answers. Focusing on a specific food group – fat, say, or fibre – does not capture the full complexity of what we eat over the course of our lifetimes, or how those different foods interact with each other and other lifestyle factors, which we'll look at in chapters to come.

Nonetheless, breaking down our complex diets into their component parts is a useful starting point for understanding the relationship between diet and health, and is the basis of official nutritional advice. To begin, let's start with perhaps the most interesting and misunderstood food group of all – fats.

THE TRUTH ABOUT FAT

For decades, dietary orthodoxy has been that fat is bad news. Not only is it the enemy of your waistline, it also clogs arteries and causes heart disease. The phrase 'a heart attack on a plate' was coined to describe the full English breakfast, swimming in grease. The idea that pigging out on such fare can lead to a heart attack is second nature to most of us; it is probably the single most influential piece of nutritional advice ever dished out.

There's no doubt that fatty food contributes to obesity – fat is the most calorie-dense of all the food groups – and being overweight is a risk factor for many diseases, including heart disease. But the idea that saturated fat is a direct cause of heart attacks appears to be melting away like a lump of lard in a hot pan.

What is a fat?

Fats are complex biomolecules that play various roles in the body, including energy storage and as components of cell membranes. A fat molecule is made up of three fatty acids bound to a molecule of glycerol. This unit is known as a triglyceride. There are dozens of different types of fatty acid, all with different properties.

The bulk of a fatty acid is a long string of carbon atoms with hydrogen atoms attached. In a saturated fatty acid, this chain does not have any carbon–carbon double bonds, meaning it has the maximum possible number of hydrogens: it is 'saturated'. Unsaturated fatty acids have at least one double bond. Fatty acids with more than one are called polyunsaturated, often used as a byword for health on food labels.

Triglycerides containing only saturated fatty acids are also called saturated; those with one or more double-bonded acids are unsaturated. As a rule, the more unsaturated a fat, the better it is for you – though this orthodoxy is being challenged. In terms of calories, however, there is no difference: saturated fats have just as much energy per gram as unsaturated fats.

Fats from animals tend to be saturated while those from vegetables are usually unsaturated. But this is only a rough guide. Meat, eggs and dairy contain unsaturated fats, while vegetables also contain saturated fats. Some vegetable fats – notably palm oil, coconut oil and the cocoa butter used in chocolate – are higher in saturated fat than beef dripping or lard.

What about cholesterol? Strictly speaking, cholesterol is not a fat. But it is lumped together with fats in the category lipids, reflecting some commonalities. Neither fat nor cholesterol dissolves in water, for example. And cholesterol is a vital link between dietary fat and heart disease. Unlike saturated fat, it is almost exclusively found in animal products: meat, fish, seafood, milk and eggs. Cutting the cholesterol in your diet doesn't have much direct effect on blood cholesterol levels but can help indirectly because cutting down on cholesterol-rich foods will usually reduce your saturated fat intake.

Saturated fat: friend or foe?

Saturated fats are found in most foods, but are especially high in meat and dairy, as well as cakes, biscuits, pastries, chocolate, avocados, palm oil and coconut oil. The idea that eating them directly raises the risk of a heart attack has been a mainstay of nutrition advice since the 1970s. Instead, we are urged to favour the unsaturated fats found in vegetables and seafood.

This advice is driven by some pretty sobering figures on the toll of cardiovascular disease (a blanket term for diseases of the heart or blood vessels, including heart attacks, strokes, heart failure and angina). According to the World Health Organization (WHO), cardiovascular disease is the world's leading cause of death, killing more than seventeen million people annually, about a third of all deaths. It predicts that by 2030, that will have risen to twenty-three million a year.

In the US the official guidance for adults is that no more than 30 per cent of total calories should come from fat, and no more than 10 per cent from saturated fat. For a man eating the recommended 2,500 calories a day, that's about as much as is in 500 grams of beef mince (12 per cent fat), 130 grams of Cheddar cheese or 55 grams of butter. UK advice on saturated fats is the same: no more than 10 per cent of total calories. That is by no means an unattainable target: an average man could eat a whole twelve-inch pepperoni pizza and still have room for an ice cream before busting the limit. Nonetheless, average adults in the UK and US manage to eat more saturated fat than recommended.

We used to eat even more. From the 1950s to the late 1970s, fat accounted for more than 40 per cent of dietary calories in the UK and US.[1] But as warnings began to circulate, people in Western nations trimmed back on foods such as butter and beef. The food industry responded, filling the shelves with low-fat cookies, cakes and spreads.

Gratifyingly, deaths from heart disease also went down. In the UK in 1961 more than half of all deaths were from coronary heart disease; now less than a third are (though cardiovascular disease is still the world's leading cause of death). But whether this is due to dietary changes is impossible to determine. Medical treatment and prevention improved dramatically, too. And even though fat consumption has gone down, obesity and its associated diseases have not.

To appreciate how saturated fat in food affects our health we need to understand how the body handles it, and how it differs from other types of fat.

When you eat fat (the triglyceride variety), it travels to the small intestine, where it is broken down into its constituent parts – fatty acids and glycerol – and absorbed into cells lining the gut. There they are bundled up with cholesterol and proteins and posted into the bloodstream. These small, spherical packages are called lipoproteins, and they allow water-insoluble fats and cholesterol (collectively known as lipids) to get to where they are needed in the body.

The more fat you eat, the higher the levels of lipoprotein in your blood. And that, according to conventional wisdom, is where the health problems start.

Lipoproteins come in two main types: high density and low

density. Low-density lipoproteins (LDLs) are often simply known as 'bad cholesterol', despite the fact that they contain more than just cholesterol. LDLs are bad because they can stick to the insides of artery walls, resulting in deposits called atherosclerotic plaques that narrow and harden the vessels, raising the risk that a blood clot could cause a blockage. This state of affairs is called atherosclerosis – colloquially and not without reason known as hardened arteries – and is the underlying cause of many cardio-vascular diseases.

Of all types of fat in the diet, saturated fats have been shown to raise bad cholesterol levels the most. Paradoxically, the amount of cholesterol you eat matters much less. The reason it has a bad name is that it is found in animal foods that also tend to be high in saturated fat.

High-density lipoproteins (HDLs), or 'good cholesterol', on the other hand, help guard against arterial plaques. Conventional wisdom has it that HDL level is raised by eating foods rich in unsaturated fats or soluble fibre, such as whole grains, fruits and vegetables. This, in a nutshell, is the lipid hypothesis, possibly the most influential idea in the history of human nutrition and a major plank of the Mediterranean diet (see page 69).

Recently, however, the consensus around saturated fat has begun to weaken – though as yet the official dietary advice has not been changed. Doubts began to creep in about a decade ago when scientists pooled the results of twenty-one dietary studies that had followed a total of nearly 350,000 people for many years. Their analysis found 'no significant evidence' in support of the idea that saturated fat raises the risk of heart disease.[2]

A few years later an even bigger analysis revisited the results of seventy-two studies involving 640,000 people in eighteen countries.[3] Again, it failed to support the status quo, and the authors concluded that 'nutritional guidelines . . . may require reappraisal'.

These doubts were reported widely, often with gusto. Many commentators interpreted them as a green light to resume pigging out on saturated fat. 'Eat Butter', declared the cover of *Time* magazine in 2014.

Can you safely ignore the old advice? For now the answer is an emphatic no. Other, less widely publicised analyses have supported the link between saturated fat and heart disease. There is also good evidence from animal research, where dietary control is possible to a degree that it is not in people. Such research repeatedly shows high saturated fat leads to elevated bad cholesterol and hardened arteries.

The results casting doubt on the orthodoxy could have arisen from other factors. It may be that in free-living humans going about their daily lives, the risk of developing heart disease depends on much more than simply the balance of saturated and unsaturated fat in the diet. Factors such as lack of exercise, alcohol intake and body weight may simply overwhelm the impact of fat.

Another key factor might be what people who cut down on saturated fat eat instead. All too often people consciously or unconsciously replace a large reduction in calories with something else. The problem is that the something else is often refined carbohydrates, especially sugars, added to foods to take the place of fat. This plays to the emerging idea that sugar is the real villain (for more on sugar, see page 20).

Then there are trans fats. Created by food chemists to replace animal fats such as lard, they are made by chemically modifying vegetable oils to make them solid at room temperature. Because they are unsaturated, and hence classed as 'healthy', the food industry piled them into products such as cakes and spreads. They also have chemical and physical properties appreciated by the food industry. They are highly resistant to rancidity and so extend the shelf life of foods. Restaurants love them because oils with trans fats can be heated and cooled repeatedly without breaking down.

However, it later turned out that trans fats cause heart disease. There is good evidence that they raise your LDL cholesterol (the bad form), and lower your HDL cholesterol (the good one), causing hardened arteries. In 2002 the US National Academy of Sciences concluded that the only safe amount of trans fat in the diet is zero.

All told, it is possible that the meta-analyses simply show that the benefits of switching away from saturated fat were cancelled out by replacing them with sugar and trans fats. But there is also emerging evidence that the impact of saturated fat and LDL is more complex than we thought.

At the moment all LDL is treated alike, but there are studies suggesting that casting it all as bad was a mistake. It is now widely accepted that LDL comes in two types – big, fluffy particles and smaller, compact ones. It is the latter that are strongly linked to heart-disease risk, while the fluffy ones appear a lot less risky. Crucially, eating saturated fat boosts fluffy LDL. What's more, there is some research suggesting small (that is, *very* bad) LDL is elevated by a low-fat, high-carbohydrate diet, especially one rich in sugars.

Why might smaller LDL particles be riskier? In their journey around the bloodstream, LDL particles bind to cells and are pulled out of circulation. The hypothesis is that smaller LDLs don't bind as easily, so remain in the blood for longer – and the longer they are there, the greater their chance of causing damage. They are also more easily converted into an oxidised form that is considered more damaging. Finally, there are simply more of them for the same overall cholesterol level. And more LDLs may equate to greater risk of arterial damage.

Complex enough? Well, there's more. Not all saturated fats are the same. A study from 2012 found that while eating lots of saturated fat from meat increased the risk of heart disease, equivalent amounts from dairy reduced it.[4] The researchers calculated that cutting calories from meaty saturated fat by just 2 per cent and replacing it with saturated fat from dairy reduces the risk of a heart attack or stroke by 25 per cent. That sounds like actionable advice, but it is far too soon to swap meat for dairy. And in any case many dairy foods – cheese especially – are high in calories and salt.

This goes back to a common problem with nutrition science. Research on single nutrients can create a misleadingly simplified picture. People do not eat saturated fat but eat foods containing mixtures of saturated, unsaturated and polyunsaturated fats, plus many other nutrients. Teasing out the effect of one nutrient within that complex buffet is very difficult.

For this reason and others it is too soon to declare saturated fat innocent of all charges; much more research is needed before the nutrition rule book can be rewritten. So while dietary libertarians may be gleefully slapping a fat steak on the griddle and lining up a cream pie with hot fudge for dessert, the dietary advice of the 1970s still stands – for now. In other words, steak and butter can be part of a healthy diet. Just don't overdo them.

THE TRUTH ABOUT OMEGA-3S

There's at least one kind of fat that most of us should probably be striving to get more of: omega-3s. These are a family of fatty acids that are vital for our health. As a key ingredient of cell membranes they have wide-ranging benefits including protecting against cardiovascular disease and cancer.

Omega-3s are usually associated with oily fish, but that is a bit of a myth. The most important one is called alpha-linolenic acid (ALA), which cannot be synthesised in the body and so must be obtained from our diet. But it is not found in fish. The best sources are chia seeds, kiwi fruit, walnuts, flax seeds (linseed), rape (canola) and soybean oil, and seaweed. Leafy green vegetables are another good source.

There are two other really important omega-3s: eicosapentaenoic acid (EPA) and docosahexaenoic acid (DHA). Both can be made from ALA but only at low efficiencies that may not supply enough. Both can also be obtained directly by eating animal products, particularly oily fish. Algae make large amounts of EPA and DHA and these fatty acids accumulate up the marine food chain, with the highest levels found in predatory fish like mackerel and tuna.

For all three, average intake among adults in the US and UK falls far short of the recommended amount, largely due to the fact that many people eat little or no oily fish. Omega-3 is probably the only nutrient deficiency that is common in the West.

Worryingly, changes in farming methods are making some fish lower in omega-3s. Half of all fish consumed globally now come from aquaculture, and farmed fish have a different nutritional profile to wild varieties. Wild salmon, for example, is an excellent source of omega-3s because it feeds on smaller fish that have eaten omega-3-rich algae. But farmed fish are increasingly fed vegetable matter, suppressing their omega-3.

Many foods are fortified with omega-3 to address consumer concerns about not getting enough (and to shift more product, of course). But for some reason omega-3-fortified foods don't seem to deliver the same benefits as foods naturally high in omega-3.

Also be sceptical of omega-3 supplements or the fish oil capsules that boast a high omega-3 content. Recent studies indicate that – unlike eating actual fish – taking these does nothing to reduce your risk of heart disease (for more on omega-3 supplements, see page 139).

Beside ALA the only other essential fatty acid is linolenic acid, which is chemically very similar. This is an omega-6 fatty acid, found in abundance in vegetable oils. Getting enough of this is not a problem. If anything, we eat too much. Excess omega-6 appears to interfere with metabolism of omega-3s, suppressing their health benefits.

Ironically, the omega-3-suppressing diet of farmed fish also increases levels of omega-6. In other words, eating too much vegetable oil and farmed oily fish may be bad for your health, which is not a message you will hear very often. But as with so many other things related to nutrition, the science is still not settled and focusing on single nutrients is likely to create problems elsewhere. The best response is to eat lots of vegetables, cut down on all fats – also a good idea for many other reasons – and try to eat wild oily fish rather than farmed.

THE TRUTH ABOUT CARBS AND SUGAR

The debate about saturated fat also touches on another food group whose reputation was seemingly sealed by twentieth-century research – albeit in the opposite direction.

The usual flip side of cutting down on saturated fat is to fill up on starchy foods. But some doctors now advocate the exact opposite: people who want to lose weight should stop worrying about fats and instead cut down on starch. If this is to be believed, it is not fat but carbohydrates we should be worried about. Potatoes, bread, pasta and rice – even the wholemeal varieties – make us fat and cause heart attacks and type 2 diabetes. Can that really be true?

What is a carbohydrate?

Starchy foods are part of a larger food group called carbohydrates, or carbs. They are a diverse bunch covering everything from simple sugars like glucose to tough, indigestible fibre. But what they have in common is that they consist almost entirely of chains of sugar molecules.

However, exactly which sugars, how many there are, and how they are linked together makes a huge difference. Carbs basically come in two flavours: simple and complex. Simple carbohydrates contain just one or two sugar molecules, such as glucose or fructose, the sugar found in fruit. Table sugar, sucrose, is a simple sugar made up of one glucose and one fructose molecule. In contrast, complex carbohydrates contain from three to hundreds of sugar units joined together. Most of the complex carbs in our diet are starches, long chains of glucose molecules linked together in a branching chain.

The carb controversy

For decades, standard dietary advice has been to fill up on complex starches, which essentially means bread, pasta, potatoes and rice. Guidelines in the UK, US and Australia, for instance, tell people to fill around a third of their plates with these foods. The recommended amount is six to eleven servings a day, more than any other food group. This advice is based on an idea from the last century that we have already encountered – the lipid hypothesis, which holds that foods rich in saturated fats are a major cause of cardiovascular disease. From the 1950s onwards, these ideas were translated into official dietary guidelines to switch to leaner cuts of meat, skimmed milk, vegetable oil-based margarines, and to fill up on starchy carbs.

Yet average body weight continued to climb, as did rates of associated problems such as type 2 diabetes. In the UK, US and Australia, around two-thirds of the population are either over-weight or obese.

In the early 2000s the orthodoxy was seriously challenged by the popularity of low-carb/high-fat diets, especially the Atkins diet, which requires you to push the pasta and rice aside and fill up on meat, butter and cream. Doctors warned it couldn't work and all that saturated fat was a heart attack waiting to happen.

And yet research showed otherwise. One trial directly compared 156 women on either the Atkins diet or a standard low-fat diet.[5] After a year, those following Atkins had lost more weight and their blood pressure and cholesterol were better than those on the low-fat diet.

Another trial put about 300 overweight women aged between twenty and fifty on either Atkins or one of three other popular diets: Zone, which cuts carbs but less severely than Atkins;

LEARN, a low-fat, high-carb diet; and Ornish, an extreme low-fat plan.

After a year, all the women had lost weight. The Atkins group lost more on average than the groups on all other diets – 4.7 kilograms versus 1.6 kilograms on Zone, 2.6 kilograms on LEARN and 2.2 kilograms on Ornish. However, only the difference between the Atkins and Zone groups was statistically significant.[6]

What might be going on? A common explanation is that fat and protein are more satiating and keep you fuller for longer, so Atkins is actually just a paradoxical way of cutting calories. There is some merit to this argument but it is not the whole story. The key factor may not be increased fat and protein, but decreased starchy carbs. In a well-meaning bid to get people to eat less fat, nutritionists may have inadvertently pushed them to eat more sugar.

One thing everyone can agree on is that excess sugar is not a healthy addition to anyone's diet (for more on added sugar, see page 26). But starch is basically made up of long chains of sugar, which are quickly broken down into actual sugars – mostly glucose – in the gut. These molecules then pass through the intestinal wall directly into the bloodstream. As far as your body is concerned, these might as well have been consumed in the form of pure sugar.

One portion of plain white rice, for example, raises your blood sugar by about as much as ten teaspoons of white table sugar. The same is true of a bowl of cornflakes (without sugar) or a hunk of baguette. The spike in blood sugar triggers the pancreas to release the hormone insulin, which causes the glucose to be taken up into cells and converted to fat. So carbs are both sugar and fat at the same time.

Even unrefined carbs, also known as wholegrain or whole-wheat, cause blood sugar to rise, albeit more slowly than their refined equivalents. A slice of wholemeal bread raises blood sugar the same amount as three teaspoons of pure sugar. A jacket potato is equivalent to eating nine teaspoons of sugar, although how fast it is released depends on what you eat with it – fat or protein lowers the rate.

Releasing insulin to manage blood sugar is a perfectly normal metabolic process but it has its limits. When too much glucose hits the bloodstream at once, it overwhelms the body's ability to deal with it. Over time, this takes a toll. The pancreas works ever harder to pump out insulin but eventually becomes exhausted. Chronic release of insulin also causes cells to become increasingly insulin resistant. Eventually, this combination of a weakened pancreas and insulin resistance can progress to type 2 diabetes.

Insulin resistance also seems to be a bigger player in heart problems than we thought. One recent study found that for men, it is a bigger heart attack risk than high blood pressure, high cholesterol or being overweight.[7]

As a rule of thumb, the more complex the carb, the better it is for you because the slower it will release its sugars. Exactly how do you figure out which carb foods are best? One pointer is the glycaemic index (GI). The GI is a way of comparing how rapidly carbohydrates affect blood glucose levels compared with pure glucose, which is given a GI of 100. Foods with a high GI (above 70), such as peeled, boiled potatoes (89) or baguettes (95), hit the bloodstream fast and cause spikes in blood glucose. Foods with a low or moderate GI (55–70) like wholegrain breads release their glucose more slowly. Hence the health halo around wholemeal bread and pasta, brown rice, bran flakes and fibrous fruit and vegetables.

GI can be deceptive, though, because it doesn't tell you the absolute amount of carbohydrate in the food. On the one hand, even low-GI foods can cause blood sugar spikes. On the other, a boiled carrot has a high GI, but contains so little sugar that it has almost no impact on blood sugar – it has a low 'glycaemic load'. Fruits, vegetables, lean meat and grains all have a low glycaemic load. So does fat. Many nutritionists now consider the glycaemic load to be the measure that matters.

So is it time to overhaul official dietary advice? Probably not. The weight of evidence is still that starchy carbs are a healthier choice than fats, though the evidence is not as solid as it once looked.

Such nuanced evidence might well leave you scratching your head over what to eat. There isn't much left if you try to avoid both fat and carbs. A more moderate approach is to avoid sat-urated fat, added sugars and refined carbs, which leaves you more or less with an extra-oily Mediterranean-style diet, high in whole grains, fish, fruit, vegetables, nuts and vegetable oils.

Another option with some evidence on its side – albeit mostly anecdotal – is a light version of the Atkins diet. Cut down on starchy food and eat lots of non-starchy vegetables and less sugary fruits such as blueberries and raspberries. In place of carbs, fill up on meat, fish, full-fat dairy products, eggs and nuts. Anecdotally, people on this diet report less hunger while also losing weight. Their blood tests show improvements in glucose control, as well as blood pressure and cholesterol levels.

That may be down to a type of carb that we have hitherto neglected. Fibre, the largely indigestible structural material found in fruit, vegetables and whole grains, slows the absorp-tion of sugars from the intestine and prevents the glucose spike. This is why healthier diets are not only low in refined carbs such as sugar, white flour and alcohol, they also contain plenty of fibre.

Poisonous potatoes?

Another reason to hold the carbs is that they may be carcinogenic, thanks to a compound called acrylamide. You may have heard the advice to avoid roast potatoes. Acrylamide is the reason for this.

As an industrial compound, acrylamide is classified as an extremely hazardous substance. The International Agency for Research on Cancer (IARC) lists it as a probable carcinogen.

Acrylamide is not added to food or found in uncooked foods. It is produced by cooking, specifically something called the Maillard reaction, which occurs between proteins and sugars when they are heated above 120 °C. The reaction produces a mixture of thousands of different chemicals that give many browned foods their appetising flavour. But acrylamide is anything but appetising. In the body, it is converted into another compound, glycidamide, which can bind to DNA and cause mutations. Animal studies clearly show that acrylamide causes all sorts of cancers.

Browning starchy foods such as potatoes produces particularly high levels of acrylamide, hence the warning about roasties. Bread is another source, especially when toasted. The chemical can also be present in breakfast cereals, biscuits and coffee.

It's hard to study the effects of acrylamide in people, but there's no reason to think that it does not damage our DNA. Quantifying the risk is difficult, however, but it probably pales in comparison to other well-known carcinogenic lifestyle factors such as smoking, obesity and alcohol. People who work in the food industry are often exposed to high levels of acrylamide, but do not have higher rates of cancer.

If you want to minimise acrylamide exposure, cut back on crisps, chips and biscuits. These are major sources of acrylamide and have the added downside of being high in sugar and/or fat. When frying, baking, toasting or roasting starchy foods, the UK Food Standard Agency's advice is to 'go for gold': aim for a golden yellow colour rather than brown. If you like your roast potatoes brown and crispy, you may have to eat them less often.

Another way of reducing exposure is to not keep raw potatoes in the fridge. At low temperatures, an enzyme called invertase breaks down the sugar sucrose into glucose and fructose, which can form acrylamide during cooking. Frozen food doesn't carry this risk, as sucrose doesn't get broken down at very low temperatures.

You can also blanch potatoes before roasting or frying them. This removes half the sugar, resulting in lower levels of acrylamide.

THE TRUTH ABOUT ADDED SUGARS

Imagine you are sitting at a table with a bag of sugar, a teaspoon and a glass of water. You open the bag and add a spoonful of sugar to the water. Then add another, and another, and another, until you have added twenty teaspoons. Would you drink the water?

Even the most sweet-toothed kid would find it unpalatably sickly. And yet that is the amount of sugar you are likely to eat today, and every day – usually without realising it.

What is added sugar?

Added sugar, or 'free sugar', refers to sugar added to food and drink (either by you or by food manufacturers) plus any sugars found in fruit juices, honey, maple syrup and so on.

The sugar added to food by manufacturers is usually either table sugar, which is sucrose, or high-fructose corn syrup. Sucrose is made up of a molecule of glucose and a molecule of fructose bonded together; they are split during digestion. High-fructose corn syrup, a mixture of glucose and fructose, is often portrayed as unhealthier than sucrose, but most researchers now agree that they are largely the same.

Calculating how much free sugar is in your diet is difficult. Food labels don't distinguish between natural and added sugar – a loophole the food industry is in no hurry to close.

Public health enemy number one?

Sugar was once a luxury ingredient reserved for special occasions. But in recent years it has become a large and growing part of our diets. If you eat processed food of any kind, it probably contains added sugar. Three-quarters of the packaged food sold in US supermarkets has had sugar added to it during manufacturing. You can find it in all sorts of unlikely places: sliced bread, salad dressings, soups, cooking sauces and many other staples. Low-fat products often contain a lot of added sugar.

It's hardly controversial to say that all this sugar is probably doing us no good. Now, though, sugar is being touted as the true villain of the piece: as bad as if not worse than fat, and the major driving force behind obesity, heart disease and type 2 diabetes. Some researchers even contend that sugar is toxic or addictive.

The WHO wants us to cut consumption radically; in 2017

it issued recommendations that adults and children should reduce their intake of 'free sugars' to less than 10 per cent of total energy intake, and preferably below 5 per cent. That would mean cutting current consumption by two-thirds, to about eight teaspoons a day for men and six for women. Many countries including the UK have introduced some form of sugar tax to incentivise people to cut down. But is sugar really that bad? Or is it all a storm in a teacup – with two sugars, please?

When nutrition scientists talk about sugar they are generally not fretting about sugars found naturally in food such as fruit and vegetables, or the lactose in milk. Instead they are worried about added sugar, usually in the form of sucrose (table sugar) or high-fructose corn syrup.

Our early ancestors would have been totally unfamiliar with these refined forms of sugar, and until relatively recently sugar was a rare and precious commodity. Only in the 1700s, after Europeans had introduced sugar cane to the New World and shackled its cultivation to slavery, did it become a regular feature of the Western diet. In 1700, the average English household consumed less than two kilograms of table sugar a year. By the end of the century that amount had quadrupled, and the upward trend has continued largely unbroken ever since. Between the early 1970s and the early 2000s, adults in the US increased their average daily calorie intake by 13 per cent, largely by eating more carbohydrates, including sugar. Today, yearly sugar consumption in the US is close to forty kilograms per person – more than twenty teaspoons a day.

The sugar rush has many causes, but one of the most important was the invention of high-fructose corn syrup (HFCS) in 1957. HFCS is a gloopy solution of glucose and fructose that is as sweet as table sugar but is typically about

30 per cent cheaper. Once this source of sweetness was available, food manufacturers added it liberally to their products. The motivation was to enhance palatability and hence increase sales.

Unfortunately, it is a guilty pleasure. Not all scientists see eye to eye on the health effects of sugar but there is one point on which most agree: we don't actually need it. You cannot live without essential fats, proteins and some carbohydrates. But sugar is an entirely dispensable food. All that unnecessary sugar adds calories to our diet, so it is no surprise that the rise in consumption coincided with the rise of obesity and related problems such as type 2 diabetes. In 1960, around one in eight US adults was obese; today more than a third are. Since 1980, obesity levels have quadrupled in the developing world to nearly one billion people. One study found that for every additional 150 calories' worth of sugar available per day in a country there is an associated 1.1 per cent rise in diabetes.

Guilty pleasure or poison?

So far, so simple. But there is a more sinister idea doing the rounds. Sugar may be more than just a source of excess calories: some forms could be directly harmful.

The object of this fear is fructose, a simple sugar found naturally in fruit but which is also a component of table sugar and HFCS.

The case against fructose is built on the fact that, unlike glucose, it doesn't play an essential role in human metabolism (that is not to say we need to eat glucose; complex forms of carbohydrate such as starch supply all the glucose our metabolisms need). Our ancestors would have encountered fructose in fruit but in nothing like the quantities we eat today, so part of the argument is that our bodies are simply not adapted to deal with large amounts of it.

To begin with, fructose is almost exclusively metabolised by the liver. When we eat a lot of it, so the argument goes, much of it is converted into fat. Fat build-up in the liver can lead to inflammation and scarring and progress to cirrhosis. Fatty liver has also been linked to insulin resistance, a precursor of diabetes.

The fructose hypothesis also says that when fructose is converted into energy it produces lots of oxygen radicals, dangerously reactive chemicals that attack our bodies. To mop these up requires antioxidants, but how many you get often depends on the quality of your diet.

What's more, unlike glucose, fructose isn't regulated by insulin. This hormone keeps blood glucose levels stable and spurs the production of leptin, the hormone that lets you know when you are full. Fructose doesn't affect leptin production and may even up the level of its counterpart, ghrelin, the hormone that makes you feel hungry. In other words, fructose encourages overeating.

Finally, eating lots of fructose has been shown in both animal and human studies to boost levels of triglycerides in the blood, which increase the risk of hardened arteries and heart disease.

But despite these claims, the case against fructose as uniquely harmful remains unproven. For now, fructose should be treated like any other sugar – with restraint. It certainly is not an argument for avoiding fruit. The benefits of a fruit-rich diet outweigh any dangers from fructose, though you should be wary of overdoing it (see five a day, page 55).

Another extreme claim against sugar is that it is addictive. For several years neuroscientists have found it useful to compare energy-dense foods to drugs such as cocaine – at least in a metaphorical sense – because it equips them with the language to discuss their habit-forming properties. But is this anything more than a metaphor?

Several studies in rats have shown that a burst of sweetness affects the reward system in the brain in a similar way to cocaine.[8] That sounds troubling, but is it also true in humans? Foods high in fat and sugar – called 'hyperpalatable' foods – are known to trigger our reward systems by boosting dopamine levels much as addictive drugs do. And there is research suggesting that most people with conditions such as binge-eating disorder display similar psychological characteristics to people with substance abuse problems. But is that enough to condemn sugar as addictive?

The scientific case for food addiction is also far from ironclad. For example, NeuroFAST, an independent, European Union-funded collaboration between thirteen universities that produces 'consensus statements' on controversial issues in nutrition science, recently reviewed all the relevant evidence from human studies. Its conclusion: there is 'no evidence' that food can be addictive.

So if we can't conclude that fructose is the culprit or that sugar is addictive, where does that leave us? Is it simply that too much sugar equals too many calories?

In a word, yes. The relationship between sugar and body weight is pretty simple. People who eat more calories gain more weight, and free sugar is an important – and avoidable – source of calories. The most important source of free sugar is sugary drinks. Why does it matter if we consume calories in liquid rather than solid form? Think of it this way. It takes about two and a half oranges to make a glass of juice. But drinking a glass doesn't make you feel as full as eating two and a half oranges. That's because the fibre in the fruit makes you feel fuller for longer. For this reason the sugar in drinks is known as 'empty calories'.

This lack of satiety from sugary drinks makes it possible to consume many more calories at a sitting than you would otherwise. Having a sugary drink with a meal, for example, doesn't make you eat less.

This lack of satiety in exchange for calories seems to have long-term consequences. Several studies have linked the consumption of sugary drinks with increased risk of obesity, type 2 diabetes and heart disease.[9] That's why soda is a prime target for public health officials.

The answer is to follow the WHO's advice to consume less than 5 per cent of your calories in the form of free sugar. That means cutting back on sugary products of all kinds, but especially drinks (for more on artificial sweeteners, see page 162).

There's a happy by-product of doing so. Everybody knows that sugar rots your teeth, and there is some evidence that cutting free sugar to 5 per cent reduces the risk. So you may not be able to eat cake, but at least you'll have something to smile about.

Of course, critics of efforts to curb sugar intake will counter that if you simply eat well and exercise, sugary drinks and snacks can be reasonable indulgences. That's true, so far as it goes. But there is also that other simple truth about sugar: however much you might want it, you really don't need it.

THE TRUTH ABOUT SALT

Sugar isn't the only crystalline white substance in our diets that gives nutritionists the shakes. Another – salt – is seen as an even bigger threat to health, and with good reason.

Salt is a routine part of most diets, casually added to food even when it is already laden with the stuff. And yet this mineral that we so enthusiastically grind and sprinkle onto food is killing us. Not immediately, but it will get you in the end. The WHO says the world is in the grip of a 'crisis' of non-infectious diseases. Salt is one of the main culprits because of its effect on blood pressure and hence cardiovascular diseases such as heart attacks and strokes. Only one substance gives the WHO greater cause for concern, and that is tobacco.

What is salt?

Salt is a vital nutrient. Its constituent ions sodium and chloride help maintain fluid balance; sodium is one of the ions nerve cells use to create electrical impulses.

The typical foods available to our hunter-gatherer ancestors would have been low in salt, so we have evolved an exquisite system for detecting it in our diet. One of our five types of taste bud is dedicated exclusively to salt, the only one tuned to a single chemical. Unlike energy, our bodies cannot readily store salt and so we are experts at hanging on to it, largely through a recycling unit in the kidneys. It is possible to survive perfectly well on very little salt.

Until recently most humans ate no salt other than what was naturally in their food, amounting to less than half a gram a day. Pure salt entered the human diet only around 5,000 years ago when the Chinese discovered it could be used to preserve food. Salt has since played a leading role in human history. It assisted the transition to settled communities and became one of the world's most valued commodities.

The rising tide

Although we no longer have to rely on salt to keep food from spoiling, our appetite for it is undiminished. Most people eat much more salt than they need. While US dietary guidelines set an adequate intake of 3.75 grams a day, the average Westerner eats about 8 grams; in some parts of Asia, 12 is the norm.

Despite a widespread belief that we have an innate taste for salt, our liking appears to be learned. People living in traditional societies, such as the highlanders of Papua New Guinea, have no access to pure salt and find it repulsive, but if they move to the city they quickly get a taste for it. As with chilli and caffeine, it seems we can learn to love the intrinsically aversive flavour. And like an addictive drug, the more you eat the more you crave, as salt receptors on the tongue become desensitised by overuse. Once in this habituated state, unsalted foods taste bland and uninteresting. It can take several weeks of salt withdrawal for taste preferences to return to normal.

It doesn't help that processed food is full of salt. Around three-quarters of the salt we eat is added to food before it even reaches our plates, not only in the obvious culprits like crisps and cured meat but also concealed in breakfast cereal, biscuits and crackers, cheese, yoghurts, cake, soup and sauces. Even bread is surprisingly salty.

There is a multitude of reasons why processed food is so laden with salt. As well as prolonging shelf life, it makes cheap ingredients taste better and masks the bitter flavours that often result from industrial cooking processes. It can be injected into meat to make it hold more water, thus allowing water to be sold for the price of meat. It improves the appearance, texture and even the smell of the final products. And it makes you thirsty, boosting sales of drinks.

This effortless consumption of salt horrifies doctors. Our kidneys can excrete some excess salt but even so, people who consistently eat more than about half a gram a day – that is, practically all of us – build up excess sodium. To keep fluid concentrations stable, our bodies retain extra water. As a result we're all carrying round up to a litre and a half of fluid, weighing one and a half kilograms, more than we would if we were on our natural salt intake.

An inevitable consequence of this excess fluid is a rise in blood pressure. Exactly why is not clear. Nor is the reason why some people are more sensitive than others. But the fact that it does raise blood pressure is uncontroversial.

It is this effect on blood pressure that causes health problems. High blood pressure is one of the main risk factors for cardiovascular disease; even small increases raise your risk of having a stroke, and everything that lowers blood pressure reduces it. For this reason, salt reduction has become one of the most important public health targets in the world. Dietary guidelines vary, but generally recommend eating no more than five to six grams of salt a day. And these levels are far from ideal – they are merely what is considered realistic in a world awash with salt. Try calculating your own salt intake and you'll soon learn how hard it is to meet even this modest target.

In theory, salt is an easy target for action. If food manufacturers slowly reduced the salt content of their products, everyone would eat less salt and nobody would even notice as their taste buds gradually re-sensitised. In the UK, this kind of salt reduction was first mooted in 1994 but hastily shelved after protests from food manufacturers. In the intervening years lobbying by scientists, public health groups and bodies such as the Food Standards Agency gradually turned the tide – not least by raising public awareness – and now the industry is broadly reconciled to modest salt reductions. Elsewhere the picture is more mixed, with US manufacturers especially truculent.

The most vigorous defender of the status quo is the Salt Institute, a trade body based in Alexandria, Virginia, representing forty-eight producers and sellers of sodium chloride. The institute has a long history of trumpeting any research that goes against the orthodoxy and picking holes in the evidence against salt. So what is that evidence? Over the years dozens of studies have been done and while the findings are far from uniform, the general direction of travel is clear.

One approach is to look for a link between how much salt people eat when left to their own devices and their rates of heart attacks and strokes. Over the years many such studies have been done; as a rule they find a clear relationship between a salty diet and cardiovascular disease.[10]

Another way is to intervene directly in people's diets – take two groups of people, force one of them to eat less salt for a while and see what the outcome is. These trials take more work than observational studies but several have been done. The biggest managed to get thousands of people to cut down on salt by about two grams a day for up to four years and saw a 25 per cent fall in cardiovascular disease.[11]

Or you can look at whole countries, taking the before and after approach. Sixty years ago northern Japan had one of the world's biggest appetites for salt – an average of eighteen grams a day per person – and shockingly high numbers of strokes. The government implemented a salt reduction programme and by the late 1960s average salt consumption had fallen by four grams a day and stroke deaths were down by 80 per cent. Finland, another salt-guzzling nation, achieved similar gains in the 1970s.

A pinch of salt?

However, there are still some uncertainties. One oft-voiced concern is that sodium may not be the whole story and that other minerals in the diet – notably potassium – also play a role in blood pressure. That is true, but in a good way. Potassium blunts the blood pressure effects of sodium, and the effect is mediated by the ratio of potassium to sodium rather than the absolute amounts.

However, modern diets are not only laden with sodium but also depleted in potassium. So an effective way to boost the health benefits of salt reduction is to switch to high-potassium table salt or eat more high-potassium foods such as bananas, oranges and black beans. However, if you have kidney disease, heart disease or diabetes, seek medical advice as an increase in potassium may be harmful.

Another perhaps more worrying possibility is that salt reduction may have unintended consequences. Big reductions in salt intake may bring about hormonal changes that also increase the risk of cardiovascular disease. This is known as the 'J-shaped curve' hypothesis because, if plotted on a graph, both extremes of salt intake are associated with an increase in mortality.

The evidence for this is still quite thin on the ground, though a recent study done in the Netherlands found that a low sodium intake of less than three grams a day was associated with an increased risk of stroke.[12] However, the researchers say that more research is needed. And in any case even if low levels of sodium are risky, you'd be hard pressed to achieve them without serious efforts to cut back on salt. Your best bet is to make those efforts.

A spicy solution

One way to cut down on salt is to use spices as a substitute. Spices have been added to national dietary guidelines in the US and Australia because adding them to food makes it easier to cut down on salt.

Spices may even bring health benefits of their own. They are often rich in polyphenols, a group of plant compounds thought to have antioxidant properties, which has led to them being touted as a kind of superfood (see page 56). The best-known super-spice is turmeric. Its magic ingredient is curcumin, which has anti-inflammatory properties. More than 120 trials have tested its effectiveness against conditions from Alzheimer's disease to erectile dysfunction, and each month dozens of new scientific papers are published on its effects.

But despite much scientific scrutiny, the verdict is far from clear. One big red flag is that however much turmeric you eat, curcumin does not get into the bloodstream. The effects may be no more than placebo.

It's a similar story with other spices. Some research suggests that cinnamon can lower insulin resistance, for example, but a review of trials concluded there was no strong evidence of any health effects. That doesn't mean spices are not good for you but a lot more evidence is needed; if they're curbing your salt intake, however, then they are certainly doing some good.

THE TRUTH ABOUT MEAT

Steak clogs your arteries; bacon causes cancer. Once thought of as the centrepiece of a balanced, healthy diet, meat is now often seen as something to be avoided. In addition to the ethical and environmental considerations that might sway your choices, meat has been linked to obesity, heart disease and cancer. But are the health warnings worth changing your diet for?

Changing our minds about meat

The first hints that meat isn't all it's cut out to be came in the 1970s when surveys started to show that countries where people ate a lot of meat had correspondingly high rates of colorectal cancer. That link was confirmed in 2007, when a World Cancer Research Fund report pulled together the results of numerous studies and concluded that red and processed meats (such as bacon, ham and sausages) were 'convincing causes of colorectal cancer'.[13] It suggested cutting out processed meat altogether and eating no more than 500 grams of red meat per week, roughly the equivalent of four burgers.

More recently, the WHO looked at the evidence and concurred that red meat – by which it means all meat from mammals, including pork – is a probable carcinogen, and processed meat a definite carcinogen.[14] Processed means salting, curing, smoking, drying or fermenting. As well as the obvious ones, this category also included jerky, biltong and canned meat. The WHO did not look at poultry (white) meat or fish, but other evidence suggests there is no link between these and cancer.[15]

The reason processed meats cause cancer appears to be added preservatives such as nitrites and nitrates, which are converted into carcinogens called N-nitroso compounds (NOCs) in the gut. However, this cancer risk needs to be put in perspective. Consider the recent finding that a bacon sandwich a day raises the likelihood of bowel cancer by 20 per cent. You could be forgiven for thinking that it means a daily bacon butty gives you a one in five chance of bowel cancer. But remember that this number is a relative risk: how much more likely you are to get bowel cancer if you regularly eat bacon sandwiches than if you do not. For an average person, the chance of getting bowel cancer at some point in their life is around 5 per cent. So a 20 per cent increase means raising the absolute risk from 5 to 6 per cent. That's big enough not to ignore, but not so big that it demands a total avoidance of bacon.

Cancer isn't the only worry. The science isn't fully settled yet, but diets high in saturated fats and cholesterol – which are found plentifully in meat – have been shown, as we have seen, to increase the risk of heart disease.

Indeed, two large studies published in 2012 found that the risk of dying from all causes – including heart disease – was 13 per cent higher for people eating 85 grams of red meat per day, and 20 per cent higher for those eating 85 grams of processed meat (which may be especially bad because it also contains high levels of salt and sugar).[16] That would translate to roughly a year off life expectancy for a forty-year-old man who eats a burger a day. That's a lot of lives potentially being shortened. In the UK, 40 per cent of men and 10 per cent of women eat a daily average of more than 90 grams of red and processed meat.

However, the validity of these conclusions is complicated by the fact that tracking exactly what people eat is notoriously tricky. For the most part researchers have to go on self-reports, which can be unreliable. People forget, under-report their intake of bad food or exaggerate the good. And as we now know, diet is intimately linked to other lifestyle factors that affect health, such as socio-economic background or alcohol consumption. Moreover, the studies vary in the way they are carried out: many don't make a distinction between different kinds of meat, for example.

Some large-scale research that does take these factors into account has found little or no connection between meat consumption and cancer or heart disease. One was a study of almost 18,000 people as part of the US National Health and Nutrition Examination Survey (NHANES).[17] It found no association between deaths from cancer or cardiovascular disease and the consumption of meat of any kind – even processed meat.

Another was the EPIC trial, which followed half a million people in ten European countries over twelve years.[18] As well as distinguishing between consumption of red, white and processed meat, it also controlled for factors such as smoking, fitness, BMI and education levels. The study found no association between fresh meat and ill health, but did see a link with processed meat. For every fifty grams of processed meat people consumed each day, their risk of early death from all causes increased by 18 per cent. The lack of any connection between unprocessed meat and heart disease came as a bit of a surprise. Their supposed heart-unhealthy effect was one of the reasons people were told to cut meat consumption in the 1970s. But recent studies hint that saturated fats aren't as bad for the heart as previously thought (see page 12).

Meat under the microscope

Fats aside, one thing meat has going for it is its nutritional value. Meat is a highly efficient source of protein, the third major food group alongside fats and carbs. Adults need around 50 grams of protein a day. To get that from meat you'd need to eat just 200 grams, or about 400 calories' worth. To get it from eggs you'd need 560 calories (9 medium eggs), from beans about 750 calories, and from nuts nearly 1,400 calories.

Meat is also a one-stop shop for essential amino acids – the ones your body needs to build proteins but can't make on its own. It is also a rich source of B vitamins, iron, zinc and other minerals and 'micronutrients'. These nutrients can be obtained from other sources, though it can be a challenge. For example, essential amino acids are found in foods such as peas and rice, but only in small quantities.

Another benefit from eating fresh meat is that it is the most readily available source of dietary iron. Ironically, though, it's the iron-rich component in unprocessed red meat, rather than its fat content, which is now generating concern. One consistent finding is that red meat is linked to cancer but white meat isn't; fish seems to be protective. The thing that makes red meat different is a compound called haem.

Haem is the iron-rich component of haemoglobin – the substance that carries oxygen around in blood – and myoglobin, which is an oxygen carrier in muscle tissue. Studies have shown that adding it to rats' food promotes tumour growth. Haem seems to produce carcinogenic molecules by oxidising fats it encounters – both in the meat and in vegetable oils.

Other problems could arise not from the meat itself, but how we cook it. Very high temperatures and contact with naked flames cause a reaction between amino acids, sugars and another meaty chemical called creatine to create carcinogenic compounds called heterocyclic amines (HCAs). Fat and other juices dripping onto hot coals, meanwhile, react to form another class of carcinogens called polycyclic aromatic hydrocarbons (PAHs). These rise in the smoke and stick to the surface of the meat. The lesson is to avoid barbecuing and pan-frying.

Nonetheless, from a health perspective at least, the weight of evidence advises against cutting out meat altogether. There's little indication that white meats, or fish, are a health concern of any sort, as long as they are not heavily processed, or pan-fried or cooked over coals. And even the health risks of red meat seem to be outweighed by the benefits. Perhaps the most surprising finding from the EPIC study was that people who ate no meat at all had a higher risk of early death from any cause than those who ate a small amount of red meat.[19] In other words, on average people who eat red meat sparingly are as healthy as vegetarians, or maybe healthier.

Why is that? Vegetarianism is not synonymous with a healthy diet. Eschewing meat doesn't exclude fatty, salty and sugary convenience foods, and many meat substitutes are heavily processed. Vegans, however, tend to be leaner than both vegetarians and carnivores.

Meat can also help people maintain or lose weight, perhaps because the high protein content makes you feel fuller and so you end up consuming fewer calories overall.

So are there ways to minimise the risks of eating red and processed meat? One answer may be fibre. The EPIC trial found that for meat eaters who reported consuming lots of fibre the risk of early death was lower than for those who ate very little meat and not much fibre either. Similarly, when people ate cold potatoes with their meat a certain kind of starch called butyrylated-resistant starch, which is produced when cooked potatoes are left to cool, seemed to protect them against the DNA damage to gut cells that is associated with colorectal cancer.

It's also possible that eating dairy alongside meat might limit the damage. France has one of the highest levels of cheese consumption in the world, yet one of the lowest levels of coronary heart disease. Some put this down to the fact that the French also consume a lot of vegetables, but several studies suggest that consuming cheese or milk causes a drop in the levels of 'bad' LDL cholesterol in people's blood. One plausible explanation is that the calcium in cheese binds to fatty acids and cholesterol in the gut, causing some of them to be excreted.

Calcium could be another way to reduce the damage caused by haem. Calcium seems to mop up haem and render it harmless; rats fed a haem-rich diet seem to be protected against its carcinogenic effects if calcium is added to their food.

And if you really can't lay off the processed meat, fruit and vegetables may help offset their carcinogenicity via chemicals called flavonoids. Indeed, concentrated flavonoids are currently being investigated as an alternative to nitrites for preserving meats. The evidence so far is that they stop microbes from growing and increase shelf life.

A good rule to live by is that you should eat no more than seventy grams of red meat per day, which is roughly a portion two or three times per week. Whether it's better to eat a little each day or to save up your quota for a weekly blowout remains unclear.

To achieve that goal you might try introducing meat-free Mondays or use meat sparingly just for flavour, rather than as a centrepiece of meals. As for how you eat meat, it seems we had it right all along: go for fresh with two veg and avoid pan-fried or chargrilled. And while you're at it, don't forget potatoes are a dish best served cold.

THE TRUTH ABOUT DAIRY

Milk is the first food we mammals ever taste. Breast milk – or synthetic versions of it – provides the perfect balance of nutrients for babies in their first year.

However, most mammals (and humans) stop consuming milk when they are weaned. But for some human children and adults, milk – usually from cows, goats and sheep – continues to be an important part of the diet.

Yet many people are voluntarily swearing off milk. Around 15 per cent of people in Europe now avoid it and other dairy products for various reasons, and the market for dairy-free substitutes is booming. For many, the souring of their relationship with milk is rooted in concerns about lactose intolerance (for more on this, see page 106) or animal welfare. But there is also a growing sense that milk is not the wholesome and healthy drink it is cracked up to be. In recent years consuming it has been linked to cancer, diabetes, asthma, acne, and even weak and fragile bones. Plant-based alternatives are often touted as healthier as well as lower in calories and fats.

In many Western countries, milk consumption has been falling since the 1970s, and dairy consumption is well below recommended levels. In 2010, the official Dietary Guidelines for Americans recommended 2.5 cup equivalents (about 0.6 litres) of dairy products per day for children aged four to eight, and 3 cups for anyone over eight. Average consumption for adults in the US is half the suggested intake.

So what's the truth about milk? Should you strive to drink the recommended amount, or avoid it altogether? If you choose to avoid it but want a substitute, what should you choose?

The skinny on milk
Milk is mostly water but also contains a complex mixture of fats, proteins and other nutrients, so nailing down its nutritional benefits and drawbacks is not easy.

One essential nutrient we get from milk is calcium. It helps build healthy teeth and bones, and plays a crucial role in blood clotting, nerve signalling and muscle contraction. Milk and dairy products are among the best sources available; without enough calcium children risk stunted bone development, and some adults are putting themselves at greater risk of osteoporosis and bone fractures, either now or in the future.

The decline of milk consumption therefore raises worries about bone development in children and adolescents, and its long-term effect on bone strength in postmenopausal women. Teenage girls, in particular, tend not to get enough calcium in their diets, which could affect bone development as they grow and increase their future risk of osteoporosis.

But milk isn't essential as a source. Calcium is also found in beans and greens – leafy vegetables, pulses – and nuts. You can also buy calcium supplements, but don't; they are generally a bad idea (see page 134). And in any case, the idea that milk is good for bone health has been challenged. Recent findings suggest that for adults, drinking too much milk might actually weaken your bones.

A few years ago scientists in Sweden followed nearly 100,000 people for many years and found that the more cow's milk people reported drinking, the more likely they were to suffer a bone fracture. The risks were greatest for women. What's more, those who consumed more milk were also more likely to die during the study; the women who drank three or more glasses a day had double the chance of dying of those who drank one.[20]

This may be due to galactose, one of the constituents of the milk sugar lactose. When laboratory animals are fed modest amounts, equivalent to one to two glasses of milk per day for humans, premature ageing and shortened life expectancy are triggered by galactose.[21] Theoretically, this could lead to bone loss and muscle damage – which would boost fracture risk – and also contribute to other diseases of ageing, which might explain the increased mortality.

Other concerns centre on the hormones in milk, such as insulin-like growth factor (IGF-1). There are claims that these hormones could boost the risk of cancer and diabetes, not to mention triggering acne and precocious puberty. One possibility is that IGF-1, plus some of the amino acids and fragments of genetic material in milk, feed into a cellular signalling pathway that takes the brakes off cell growth, boosting the risk of cancer and other diseases. Constantly high levels of IGF-1 could also cause insulin-producing beta cells in the pancreas to burn out, resulting in type 2 diabetes. But these ideas are controversial.

Reviews by both the World Cancer Research Fund and the American Institute for Cancer Research have concluded that while there is a probable link between dairy product consumption and prostate cancer, milk and dairy also probably protect against colorectal cancer and may lower the risk of bladder cancer. The evidence for this protective effect is stronger than for the causation of prostate cancer.[22]

So why is it so difficult to get a clear-cut answer on the health effects of milk? One reason may be that milk itself is so complex. It promotes IGF-1 production, and it contains whey proteins and protein fragments called peptides, all of which, like calcium, have signalling roles, sending messages between cells that modulate metabolism. A protein called lactoferrin, for instance, plays a crucial role in iron absorption and exerts anti-oxidant and anti-carcinogenic effects. Meanwhile, peptides derived from another milk protein called casein may influence the behaviour of immune cells and prevent harmful bacteria from attaching to the gut.

Then there's the question of what we consume alongside milk as part of our complex diet. One reason why milk and other dairy products may protect against colorectal cancer is because calcium seems to bind to and neutralise carcinogenic substances in meat.

This interplay with other aspects of our diet could also benefit heart health. High consumers of milk and dairy have a 16 per cent lower risk of heart disease and stroke than low consumers.[23] Again, the reason could be calcium, which binds to dietary fats in the intestines and also to some of the bile acids that help to digest fat, preventing its absorption by the body.

A common problem is that studies rarely distinguish between milk, cream and butter on the one hand and fermented products like yoghurt, soured cream and cheese on the other. From a nutritional perspective butter and cream are essentially concentrated milk; they contain the same basic mixture of ingredients, including lactose (and hence galactose). Fermented products are somewhat different because the lactose has been broken down, but in other respects can also be thought of as concentrated milk – though many cheeses also have high levels of added salt.

Cheese contains around six times as much calcium weight for weight as whole milk (as a rule it takes ten kilograms of milk to make one kilogram of cheese). The increased risk of fractures and mortality associated with milk consumption is not seen in people who eat a lot of fermented milk products. If anything, the opposite is true. While there is not enough evidence to change official dietary advice as yet, these findings suggest it might be wise to get some of the protein, vitamins and minerals milk provides from other healthy sources, such as yoghurt, nuts and seeds.

These aren't particularly pourable though, so what are your options if you want a white coffee or a bowl of cereal?

Milking it
A growing number of products derived from milk are now on the dairy market, but are any of them worth it?

- **LOW FAT MILK** such as skimmed and semi-skimmed is popular, but there has been little research into whether it offers any health benefits over the full-fat variety. Recent studies have hinted that some saturated fats may in fact be beneficial, and that full fat milk could help with weight loss.
- **LACTOSE-FREE** is simply regular milk that has had the enzyme lactase added to it to break down the milk sugar lactose. It enables people who are lactose intolerant to drink milk. But in nutritional terms it is the same as regular milk.
- **ORGANIC MILK** comes from cows that are allowed out to graze whenever conditions allow. This means it contains higher levels of omega-3 fatty acids, which are found in grass. Even so, the total amount it contains is small in relation to the whole of a person's diet.

  Another common reason to go organic is fear about hormone levels in non-organic milk. All milk naturally contains hormones, but in areas where cows are treated with growth hormones – as happens in some US states but not in the European Union, Canada, Australia or New Zealand – non-organic milk may have higher levels of insulin-like growth factor, a hormone linked to increased risk of some health problems. But the US Food and Drug Administration concluded that it poses no health risk at the levels present in mass-produced milk.

- **A2 MILK** does not contain a protein called A1 that is a form of one of the most abundant proteins in cow's milk, beta-casein. This comes in two forms, A1 and A2. They only differ by a single amino acid, but this influences how they are digested in the gut. The breakdown of the A1 type can form a peptide called beta-casomorphin-7, which has been claimed to increase the risk of diabetes and heart disease. However, a review by the European Food Standards Agency concluded that current evidence doesn't support the idea that A2 milk is healthier.

- **UNPASTEURISED (OR 'RAW') MILK** can be bought direct from farmers, at markets or through a delivery service in some countries; in others it is banned because of fears about food poisoning. Proponents of raw milk claim that it tastes better, guards against certain health conditions and is easier for people with lactose intolerance to cope with, but evidence fails to back this up. A recent review concluded that salmonella and E. coli represented a genuine threat to consumers of raw milk. In the days before pasteurisation, an estimated 25 per cent of all food and waterborne disease outbreaks in the US were associated with milk, which is an ideal growth medium for microbes.

Don't have a cow

If real milk is not your cup of tea, there are a growing number of plant-based alternatives, also known as alt-milks. Alt-milks are often perceived as healthier than cow's milk, but in reality their nutritional properties vary enormously. They are made by grinding plant material, adding water to make a slurry and then straining it. So, unsurprisingly, their nutritional properties depend

a lot on the starting material. Almost all of them are also fortified with vitamin D, vitamin B12 and calcium.

The market is dominated by almond, soy and coconut milks
but you can also buy alt-milks made from macadamia nuts,
cashews, peanuts, rice, oats, peas, flax, hemp, bananas and even
potatoes.

Apart from soya milk, most of them are low in protein, some
extremely so: rice milk has just 0.1 per cent protein. Cow's milk
is just over 3 per cent protein. Soya milk is about the same, and
it contains similar amounts of the omega-3 fatty acids that are
important for heart health.

Plant-based milks also tend to be low in calcium, unless they
are fortified. Even then there is evidence that this doesn't confer
the same health benefits as calcium from milk. The calcium in
dairy products is similar to the calcium compounds in the body
so it's more readily incorporated into the bones.

Plant-based milks may have other things going for them,
however. Some are low in calories, which could be helpful for
weight control. Almond and cashew milks, for example, have
less than half the calories found in cow's milk.

Coconut and hemp milk have a high fat content but also
include a small amount of dietary fibre not found in cow's milk.
Oat and rice milks are higher in carbohydrates than both cow's
milk and other alternatives; rice milk often has significant
amounts of sweeteners added to improve the flavour.

Any one of them can be considered healthy when combined
with a balanced diet, though the same can be said for dairy
milk. To a large extent, it is swings and roundabouts.

THE TRUTH ABOUT BREAD

Wheat was one of the first crops to be domesticated and for thousands of years it has been a staple food for much of the world. Bread is deeply ingrained in many cultures and is highly evocative of homeliness and goodness – hence the popular advice to bake a loaf of bread just before prospective buyers come to view your house.

But wheat has lately come under attack from people who claim that it causes all sorts of uncomfortable symptoms, from bloating to headaches, joint pain, fatigue and obesity. Celebrities and sporting heroes have promoted the wheat-free lifestyle, notably tennis superstar Novak Djokovic, who credited it with turning him from a fine but fragile player into a ruthless winning machine. Testimonials like Djokovic's have helped to cement the idea that there is something wrong with eating wheat, and that cutting it from our diets can have almost miraculous results.

This identification of vague but wide-ranging symptoms with a single, simple culprit has led to an epidemic of self-diagnosed 'gluten intolerance'. We are told that around one in five people would benefit from cutting wheat out of their diet to some degree. Around a third of Americans are reported to be considering it. Demand for gluten-free bread, pasta, noodles and even beer is rising rapidly.

We'll go into more detail on food intolerances in the next chapter, but it is worth spending some time on the nutrient in bread that reaps the blame: the protein called gluten.

Gluten is actually a complex of two proteins found naturally in wheat, glutenin and gliadin, which form a network of fine stretchy strands when wet. This is what gives dough its elasticity and makes wheat so versatile, allowing it to be turned into everything from bread and pastry to pasta and noodles. Gluten is found in other grains besides wheat, including barley, rye, oats and some rarer grains such as spelt and emmer.

There is no doubt that some people – but only a tiny minority – must avoid gluten or risk serious health problems. There is also some evidence that for some people 'gluten intolerance' is a real thing (for much more detail on this, see page 102). However, the vast majority of people can deal with gluten, no problem; as far as your digestion is concerned it is just another protein among many.

Nonetheless, the gluten bandwagon rolls on. People with gut problems are far from the only ones going gluten free. About a third of the US population say they would like to cut down or eliminate gluten from their diets. Surveys show that many are choosing to avoid gluten because they believe it will help them lose weight or improve their overall health.

The evidence for many of these claims is thin or non-existent, and many nutritionists see the gluten-free craze as a triumph of marketing over science. Wheat is undoubtedly fattening if you eat too much of it, but so are most foods. It's hardly surprising that people who cut it from their diets can start to lose weight. After all, a no-wheat diet means putting down many high-calorie foods such as cakes and biscuits, as well as swearing off the beer. Being careful what you eat will pay dividends whatever your diet.

But be aware that holding back the wheat can have nutritional downsides. Ditching cakes and biscuits is probably a good idea but swapping wholewheat bread for processed, gluten-free substitutes is probably not. People with coeliac disease are, for example, warned of the risk that they could be deficient in key nutrients, as their diets generally contain more fat and less fibre than those of people who eat gluten. That is bad because eating a high-fibre diet curbs your appetite, and it also reduces your chances of developing several cancers. And wheat, particularly wholegrain wheat, is a major source of fibre.

So eating gluten-free foods is not necessarily less fattening, nor is it even healthier in general. In general they are low in fibre and have a high glycaemic index, which means they are digested quickly and cause unhealthy spikes in blood sugar.

Perhaps a more appetising option is to try eating better bread. Humans have been making bread for 10,000 years, but the way we do it has changed dramatically in the past half-century. In 1961, a new method of mass-production was invented at the British Baking Industries Research Association in Chorleywood, near London. It used extra yeast, additives and industrial mixers to speed up fermentation, so a loaf could be made in just a few hours rather than days. It also meant that lower-quality flour could be used for bread. Around 80 per cent of bread in the UK is now made this way, and the Chorleywood Process is used to some extent in many other countries.

But there are concerns that this short cut alters the digestibility of bread, which may explain why many people name bread as a trigger for bowel problems such as bloating and excessive gas.

Experiments have been done comparing the effects of fast- and slow-fermented breads on gut microbes. Fast-fermented bread produced more gas than sourdough, which is left to rise for several hours using natural yeast. The implication is that if bread is left to ferment for longer, its carbohydrates will reach the gut in a pre-digested state and gut bacteria won't have as much to work with. So before slicing it out of your diet, use your loaf and first try buying better bread.

THE TRUTH ABOUT YOUR FIVE A DAY

Scientists have known for a long time that people whose diets are rich in fruits and vegetables are less likely to have heart disease, diabetes, dementia, stroke and certain types of cancer, and (unsurprisingly) tend to live longer. The reason for this is no mystery: fruit and veg contain high levels of vitamins, minerals, antioxidants, fibre and other healthy compounds, plus low levels of fat and salt. Official advice in the UK is to eat five portions a day.

Such advice is certainly based on sound science. According to the WHO there is convincing evidence that diets featuring lots of fruit and veg lower the risk of heart disease and probably cancers. They also seem to promote better mental health.

But for some reason, achieving five a day is often portrayed as a great challenge, as if fruit and vegetables are an unwelcome addition to our natural diet rather than the centrepiece of it. Admittedly some people don't like them very much, but there's such a range of options that disliking, say, green vegetables shouldn't be a barrier to hitting this most simple and beneficial of targets.

One problem is in knowing what constitutes one of the five. Simple: a portion of fresh fruit or veg is 80 grams, equivalent to an apple, two tangerines or seven cherry tomatoes. Dried fruit also counts; a portion is 30 grams, about a handful of banana chips (though beware, as these are often sugar coated). Potatoes don't count, but baked beans do – though only up to a maximum of one portion a day, as per all pulses and beans.

It's worth bearing in mind that, nutritionally speaking, unprocessed frozen fruits and vegetables are almost as good as fresh ones. But that's less true of processed short cuts such as fruit juice, dried fruits or yoghurts with fruit pieces in them (a glass of fruit juice counts but, again, only up to a maximum of one portion a day; any more is just sugar). As a general rule the more processed the fruit, the less it counts.

Targets are higher in some other countries. The US, for example, recommends eight to ten a day. And the science suggests this is right. A recent review of ninety-five studies of the relationship between diet and health found that the ideal intake is ten portions a day.[24] People who ate that amount had nearly a third lower risk of death than those who ate none during the course of the studies, which followed people for between three and thirty years. Most of the benefits stemmed from reductions in the rates of heart disease and cancer, the commonest causes of death in Western countries.

If you already struggle to eat five a day don't despair. More health benefits are accrued from going from zero to five than from going from five to ten.

But if you do get over five, be careful how you hit the target. Ten portions of fruit contain a lot of sugar, and fruit sugar is no healthier than the sugar in sweets. The best advice is to eat a wide range of fruit and vegetables, which will also maximise the health benefits. Risk of cancer, for instance, falls as people eat more cruciferous vegetables such as cabbage, sprouts, broccoli and other green vegetables. The old adage is right: eat your greens.

THE TRUTH ABOUT SUPERFOODS

You don't have to shop at a whole food store to have heard rumours about exotic seeds that replenish your energy or berries from the other side of the world that disease-proof your organs. Such are the promises of so-called superfoods, an ever-expanding category credited with all manner of miraculous properties – often with a price tag to match.

If it is all hype, we are swallowing it. In a survey of more than 1,000 UK adults, 61 per cent admitted to buying a food because they considered it a superfood, 30 per cent agreed that 'superfoods are scientifically proven to have health benefits' and 14 per cent said they were willing to pay more as a result. Are we being ripped off?

What is a superfood?

The first thing is that the term 'superfood' has no scientific meaning. The Oxford English Dictionary defines one as 'a nutrient-rich food considered to be especially beneficial for health and well-being', but the term was invented to sell products. Yep: superfoods are a marketing gimmick.

In 2007, the European Union effectively banned the word on packaging. There are no such regulations in the US, but the Food and Drug Administration can take action if they find any claims to be false or misleading. Even where claims are not directly misleading, the scientific jargon designed to attract health-conscious consumers is murky. Sure, studies often show that concentrated extracts or isolated compounds found in abundance in certain foods have particular effects in a Petri dish or a mouse. But that does not mean they work in the same way when people eat them. Reliable, long-term studies to support most claims are thin on the ground.

The most eye-catching health claims for superfoods tend to be associated with particular groups of compounds: the glucosinolates in kale, for instance, or the anthocyanins in blueberries. Trouble is, although scientists can study how isolated compounds act on cells in a dish or in mice, it's much trickier to get a grip on what foods containing these ingredients do inside the human body.

The way to do experiments is almost like a drug trial, testing one nutrient against something that doesn't contain that nutrient. But we don't eat nutrients in isolation. Nor do we have a great understanding of how the foods we eat are broken down into their molecular building blocks, how these interact with each other and what happens to them when, or if, they reach different tissues. There is also variation from person to person in the gut bacteria that influence all of these processes. And in some cases, cooking can alter the levels of active ingredients.

Even when scientists do put superfoods to the test in human studies, it's hard to figure out how they perform. Hence most claims you hear refer to cell or mouse studies and should be taken with a pinch of salt.

Super or not so super?

For the average consumer, it can all get a bit confusing. So which foods, if any, are really super?

Kale

Kale used to be the vegetable you'd fall back on if there was an unexpected shortage of cabbage, but it is now one of the brightest stars of the superfood firmament, blended in smoothies, baked to crisps, blanched and dressed with olive oil or served raw in salads. Its main selling point is the family of sulphur-containing plant chemicals, glucosinolates, found in abundance in all dark-green cabbage-like vegetables (variously called brassicas, mustards and crucifers) and impart their characteristic bitter taste.

Glucosinolates are broken down in the gut to release glucose and another chemical mouthful called isothiocyanates, which have been shown to stimulate enzymes whose job it is to eliminate cancer-causing chemicals. This much has been shown in animal research and it is supported by studies linking higher consumption of glucosinolate-rich vegetables to a lowered risk of cancer in humans, particularly that of the gut and lungs.[25]

However, all brassicas contain similar compounds in similar amounts, so kale is no better than, say, broccoli or cabbage.

What's more, there are more than a hundred different glucosinolates, and each gets broken down into a different isothiocyanate, so it is probably best to eat a wide range of brassicas rather than just kale. Be warned, though: boiling reduces glucosinolate content, so al dente is better.

Brassicas also contain other healthful compounds such as polyphenols, antioxidants and vitamin C, which may be why high intake has been associated with a decreased risk of chronic diseases. These compounds are found in especially large amounts in traditional varieties such as cavolo nero and grelo/rapini.

VERDICT: Super, but no more so than other brassicas.

Quinoa

Originally from the Andes, quinoa is the quintessential superfood. Its seeds are rich in protein, fibre, vitamins and minerals, and cost a packet. The fact that they are also gluten free only adds to the halo.

A handful of studies have shown that replacing cereals with quinoa can reduce blood cholesterol and help with weight loss.[26] But the sample sizes are so small that it is difficult to draw firm conclusions.

Such supposed benefits are typically attributed to chemicals called saponins, thought to work by altering the permeability of the gut. It may also be because they produce a bitter taste that puts people off eating too much. Either way, washing your quinoa before eating it, as many people do, removes the saponins and thus any benefits they might bestow.

Despite the hype we don't know much for definite. So although quinoa is a perfectly healthy addition to your diet, there is no compelling health reason to replace staples like rice or wheat.

VERDICT: Eat if you like it, not for health benefits.

Berries

Berries – especially blue ones – are prized for their ability to lower the risk of cardiovascular diseases. A study of 93,000 women, for instance, showed that participants who ate three or more portions of blueberries and strawberries a week had a 32 per cent lower risk of a heart attack than those who ate berries once a month or less.[27]

Such benefits are typically attributed to a compound called anthocyanin, part of a family of plant chemicals called flavonoids. These are found in particularly high levels in blueberries and red berries such as strawberries and raspberries. Which sounds great, but there's a catch: hardly any of the anthocyanins in berries get into the bloodstream.

It's possible that it is not the anthocyanins that protect your heart, but the chemicals they are broken down into. These get into the bloodstream in far higher quantities than the original compounds.

There could be other explanations. Maybe some other component stimulates your own defences. Or perhaps anthocyanins act as benevolent gardeners for your colon microbiome, nudging it towards a healthier mix.

The berry most commonly purported to have superpowers is the goji. Long revered in Chinese medicine, goji berries (*Lycium barbarum* and *Lycium chinense*) are said to pep up your immune system and protect against heart disease and cancer. But there is precious little research to back these claims. The other big boast is that goji berries contain high levels of zeaxanthin, a compound linked to the prevention of age-related degeneration in eye cells. That is true, but many other foods do too, for less money. If it's zeaxanthin you're after, you can get your fill from spinach, cabbage or yellow peppers. And although goji berries do contain more vitamin C than blueberries, you get roughly the same amount from strawberries or lemons.

VERDICT: Super-ish.

Chocolate

It would be great if chocolate were a superfood, but the evidence is flaky. The most commonly heard health claims centre on chemicals called flavanols, the dark and bitter compounds found in cocoa beans. But long-term trials are lacking, and you'd have to eat heroic amounts of chocolate to benefit. One study on the memory-enhancing effect of flavanols found a positive association, but only after eating half a kilo of dark chocolate a day. And in any case, much of the research is funded by the chocolate industry.

What's more, any benefits of cocoa flavanols have to be weighed against the fact that most chocolate contains large amounts of sugar and fat. That's not to say that chocolate, especially the expensive cocoa-rich stuff, is something you should avoid – just don't fool yourself that it is a health food.

VERDICT: Fine occasionally, but no excuse to gorge.

Chia seeds

The ancient Maya used chia seeds for thousands of years to make everything from flour to tea, but don't take this as some sort of ancient wisdom. They did it out of necessity. Now chia seeds are one of the trendiest superfoods, often said to be a great source of omega-3s.

Chia seeds contain roughly 17 grams of omega-3 per 100-gram serving. Compared with the go-to source of omega-3, oily fish, that looks impressive: a 100-gram portion of farmed Atlantic salmon only gives you 2.2 grams of omega-3s. Unlike salmon, however, the fatty acids in chia seeds come as alpha-linolenic acid (ALA), which the body has to convert to eicosapentaenoic acid (EPA) and docosahexaenoic acid (DHA) to get the full health benefits.

Human metabolism doesn't do this very well. It varies from person to person, but the average efficiency is below 10 per cent, so the amount of useful omega-3s you get from 100 grams of chia seeds is just 1.7 grams, less than from salmon. You also have to consider that to convert ALA, your body has to digest the seeds to extract the fats – and some seeds pass straight through.

VERDICT: Good, but oily fish packs more omega-3s.

Kimchi and kefir

If you've somehow missed them, kimchi is Korea's national dish of fermented cabbage, and kefir is a fermented milk drink from the Caucasus in eastern Europe. These and other fermented foods have been around for centuries but are now acclaimed as microbiome-balancing paragons of dietary virtue.

Some of their powers are put down to the fermentation process, in which bacteria partially digest the food, releasing a greater hit of nutrients. Fermentation does seem to improve the availability of iron: one recent study in humans showed greater absorption of iron from fermented mixed vegetables than fresh ones.[28]

The biggest health claim, however, is the supposed effects on your gut microbiome, the billions of bacteria that occupy your intestines and are involved in many vital metabolic functions. The idea is that the fermentation process increases the numbers of beneficial bacteria naturally present in the food and, by extension, in your gut.

Animal studies suggest that fermented foods might encourage a healthy mix of microbes in the gut.[29] Adding kefir to the diets of mice, for instance, increased the population of beneficial bacteria like Lactobacillus and Bifidobacterium and reduced potentially harmful ones.

However, caution is needed when it comes to the benefits for humans. Fermented foods do look as if they could benefit gut flora, but what this does for your health is not clear.

VERDICT: May be good for gut bacteria.

Wheatgrass

Fans of the wheatgrass shot, the dark-green sludge liquidised from the young shoots of wheat, insist that it will flood your tissues with oxygen. The story goes that chlorophyll, the compound that plants use to make sugars via photosynthesis, is structurally similar to haemoglobin. So as there is more chlorophyll in wheatgrass shoots than in other edible plants, you get more oxygen.

Nonsense on stilts. Chlorophyll is found in all green vegetables. More to the point, there is no evidence that chlorophyll functions anything like haemoglobin. Even if it did, it wouldn't get into the bloodstream because chlorophyll gets broken down in the gut.

VERDICT: Bunkum.

Chillis

If you like it hot, here's some good news. When scientists in China followed the diets of 480,000 apparently healthy adults for seven years, those who ate chilli almost every day were 14 per cent less likely to die over the course of the study than those who ate it less than once a week.[30]

That may be because of chilli's effect on energy metabolism. The active ingredient in hot peppers is the heat-producing compound capsaicin, which may activate a body tissue called brown fat. Whereas the more common white fat stores energy, brown fat converts food directly into body heat. Cold triggers brown fat to turn up the thermostat, converting calories into heat and promoting weight loss. Capsaicin seems to mimic this effect, which makes sense to anyone who has had the chilli sweat after a hot curry (though any weight lost this way will probably be overwhelmed by the calories in your food).

As usual, however, there's a catch. Brown fat probably evolved to keep babies and infants warm. It starts to degenerate during adolescence and most adults don't have any left.

But chillies are rich in polyphenols so there's still gain to be had from the pain. Just don't assume they are superior to other vegetables.

VERDICT: Hot, but not super.

Beetroot

Once a preserve of long-forgotten jars marooned at the back of the fridge, beetroot is now lauded as a blood-pressure lowering, metabolism-revving superfood. The main goodies are said to be nitrates. These are converted into nitrites by saliva and then pass through the stomach, where they are converted into nitric oxides – compounds that relax blood vessels.

Indeed, studies have shown that dietary nitrate brings down blood pressure and improves circulation. It might even improve physical endurance: one study found that 500 millilitres of beetroot juice per day improves exercise performance, buying an extra 90 seconds of intense exercise before exhaustion.

Don't overdo it, though. The European Food Safety Authority (EFSA) puts the safe limit for dietary nitrate intake at around 260 milligrams per day for an average adult – that's equivalent to 300 millilitres of beetroot juice or two whole beets. The main concern is that if nitrites combine with protein in the stomach, they could potentially form nitrosamines, which may contribute to cancers. The link is not proven, though, and adding vitamin C may prevent the formation of nitrosamines.

On balance, nitrates are a good addition to your diet and beetroot is a great source. As ever, though, there are plenty of alternatives: lettuce, rocket and other leafy vegetables are also good sources.

VERDICT: Good stuff, but don't beet it too hard.

Avocado

If you factor out the risk of 'avocado hand' – an injury caused by a slip of the knife while preparing avocados, which is apparently on the increase as the popularity of the fruit rockets – then eating avos appears to be good for your health.

A study of more than 55,000 adults found that avocado eaters were slimmer than abstainers.[31] But the results were obtained from participants' self-reports, a notoriously unreliable method.

VERDICT: Eating fruit and vegetables is good for you, but we knew that already.

THE TRUTH ABOUT ORGANIC FOOD

Health and safety usually get a bad rap, but when it comes to food we can't get enough of it. Who wouldn't want to eat food that is healthier and safer? That is one of the reasons why organic food is so popular (at least from the personal health perspective; another appeal is that it claims to be more environmentally benign with better animal welfare standards). Despite being more expensive than non-organic, sales of organic fruit, vegetables, meat and dairy have been rising steadily for years.

As with many nutritional claims, the idea that organic equals health appeals to common sense. Conventionally produced fruit and veg are doused in pesticides and fertilisers; intensively farmed animals are fed processed food and pumped full of hormones and antibiotics. Surely raising crops and animals in a more natural system makes for more nutritious food? You'd think so, but common sense is no substitute for sound science.

Organic agriculture first became popular among consumers in the 1970s but the movement began about a hundred years ago, around the same time as the invention of industrialised agriculture. Definitions of what qualifies as 'organic' vary from place to place and can be enormously technical, but as a rule of thumb organic farming strictly limits the use of synthetic chemicals and instead relies on organic fertilisers such as manure and bone meal, and naturally occurring pesticides such as pyrethrins (derived from chrysanthemums) and Bt (from bacteria). Livestock farmers do not use feed additives or artificial growth promoters such as hormones and antibiotics, but can use drugs to treat sick animals. Organic farming is thus more natural but, contrary to popular belief, not completely free of added chemicals. This is not a misconception that the organic industry is straining to dispel.

As a consequence you might assume that the food is healthier, and advocates of organic food often passionately believe this to be the case. But the scientific evidence is as thin as the topsoil on an over-grazed field.

Consider pesticide residues, for example, which are an important motivation for buying organic food. Synthetic pesticides are definitely not good for you. But there are strict consumer safety laws governing how much can be left on produce.

The short- and long-term health effects of consuming the residues, even in legal quantities, are not known, and eating organic food will definitely reduce your exposure to them. Three pesticides used regularly in conventional agriculture – glyphosate, malathion and diazinon – are classed as probable human carcinogens, and people who switch from conventional to organic produce have lower levels of all three in their urine.[32] However, the relevance of these findings to human health is not clear. If you think it is a good idea to minimise your exposure to chemicals that are classed as probable carcinogens, organic produce will help.

However, there are other considerations. Recent research found that a bigger worry is contamination with pollutants such as PCBs and heavy metals, and long-banned but highly persistent pesticides such as DDT and dieldrin.[33] For these it matters less whether the food was grown organically or non-organically than whether it was grown close to a source of the pollution. An organic label is no guarantee that it is free of nasty chemicals.

What about claims that the food is more nutritious? Again, the science is scarce and often contradictory. But on balance, organic comes out slightly ahead. A recent major review of the evidence concluded that people following an organic diet were healthier overall and less likely to develop chronic diseases.[34] The authors speculated that this might be due to the food's higher nutritional value, or a lack of pesticides. But as with all things in nutrition science, proof is hard to come by. It may simply be a confounder: people who buy organic food tend to be wealthier and lead healthier lifestyles in other ways.

One foodstuff where the data is quite strong is milk. Recent UK and US studies found that organic milk from cows reared outdoors had higher amounts of healthful antioxidants and omega-3s.[35] The difference is down to the cows' diet, as cattle on organic farms have greater access to fresh grass, which supplies the building blocks for these compounds in the milk.

If health is your goal, buying organic food certainly won't do you any more harm than conventional food. But its environmental credentials have been questioned and it costs more, so you might be better off spending the money on something with more evidence behind it, such as a gym membership (as long as you use it).

THE TRUTH ABOUT A HEALTHY DIET

So what do we know about what actually constitutes a healthy diet? If you had asked this question ten years ago, the answer would have come back loud and clear: plenty of fresh fruit and vegetables, grains and nuts, a bit of fish and poultry, olive oil instead of butter, a glass of wine with your food – but go easy on the red meat, dairy and refined sugar. For decades this

'Mediterranean diet' was lauded as the recipe for a long and healthy life. Dietary guidelines in the UK, the US and elsewhere reflect its key messages, especially on meat and dairy, and millions of people try to stick to it in the belief that it has been scientifically proven.

Recently, though, this sunshine diet has fallen under a bit of a cloud. Several studies have called its benefits into question, and the original research into its health-giving properties has been heavily criticised. As we have seen, many of its individual components – especially the avoidance of fat – have been challenged by new evidence. So is it time to rewrite the menu?

To answer this question it helps to know a bit of nutritional history. The idea that there was something special about the traditional diets of countries ringing the Mediterranean Sea first took hold in the 1950s, when a rising tide of heart attacks among middle-aged men was spreading alarm in the US. At the time it was explained as an inevitable consequence of ageing. But American nutritionist Ancel Keys of the University of Minnesota begged to differ.

Keys noted that heart attacks were rare in some Mediterranean countries and in Japan, both places where people ate a low-fat diet. Convinced that there was a causal link, he launched the Seven Countries Study in 1958. It was a Herculean effort. In all, he recruited 12,763 men aged forty to fifty-nine in the US, Finland, The Netherlands, Italy, Yugoslavia, Greece and Japan. Their diet and heart health were checked five and ten years after enrolling.

The groups were chosen to represent a spectrum of diet and heart disease incidence. The Netherlands and Greece were included because they were thought to have high-fat diets with low levels of heart disease. The US and Japan represented the high and low extremes of fat consumption.

After poring over the huge data set this produced, Keys concluded that there was a clear association between saturated fat intake and incidence of cardiovascular disease. This became known as the lipid hypothesis, which has been called 'possibly the most influential idea in the history of human nutrition'.

The finding was supported by other research, notably a big US study that tracked diet and heart health in the Massachusetts town of Framingham. In light of this research and the rising death toll – by the 1980s nearly a million Americans a year were dying from heart attacks – health authorities decided to push for a reduction in fat, and saturated fat in particular. Official guidelines first appeared in 1980 in the US and 1991 in the UK.

More recently, however, well-publicised doubts about Keys's conclusion have begun to circulate. One common complaint is that he cherry-picked data to support his pre-existing views about fat, ignoring countries such as France that had high-fat diets but low rates of heart disease. The strongest evidence in favour of a low-fat diet came from Crete, but it transpired that Keys had recorded some food intake data there during Lent, a time when Cretans traditionally avoid meat and cheese, so their diets appeared to be less fatty than they actually were.

The Framingham research has its detractors too. Critics say that it followed an unrepresentative group of predominantly white men who were at high risk for heart disease for non-dietary reasons such as smoking.

There have also been problems with modern research. In 2013 the results of a major study of the benefits of the Mediterranean diet were published.[36] The PREDIMED trial

(Prevención con Dieta Mediterránea) followed more than 7,000 people in Spain for nearly five years; it confirmed that the diet lowered the risk of cardiovascular disease, reduced blood pressure, prevented heart attack and slashed the risk of stroke. But a few years later these results were called into question due to problems with the study design. The flagship research paper was actually retracted, meaning that scientists no longer accept its findings.

So is it time to revise our view of the Mediterranean diet as the closest thing we have to an ideal? If you're hoping that the answer is yes, and that fat and sugar and red meat are back on your plate, you're in for a disappointment. The fine details probably need a rethink, but in broad-brush terms a Mediterranean style still holds up.

For example, the PREDIMED researchers reanalysed the results to correct for their errors and found that the protective effects of the Mediterranean diet held up, but only for people with a high risk of heart disease.[37]

Another recent study suggests that the Mediterranean diet works, but you can't do it on the cheap.[38] A four-year trial of 18,000 people in Italy found a 15 per cent reduction in cardio-vascular risk among those following the diet, but only for people earning more than £35,000 a year. For the less well off, the benefits weren't seen at all.

The difference turned out to be in the quality of their food. Poorer people ate less fruit, vegetables and whole grains, bought cheaper olive oil and subsisted on foods that had lower nutritional value. Which is ironic. When the Mediterranean diet was discovered, it was the diet of the poor. Now, it's the diet of the rich.

All in all, the Mediterranean diet is holding up quite well. Nutrition research is, by its very nature, difficult and imprecise, as we'll see in the next section. But a lot of research supports the idea that the healthiest diets we know of feature lots of fruit and vegetables, and limited saturated fat. In 2014, Italian scientists published a round-up of traditional Mediterranean-style eating patterns. 'All findings pointed to the same conclusions, that is, the protective role of dietary habits characterised by a predominance of plant foods and fish over animal foods and sugar', the researchers concluded.[39]

Meanwhile, although obesity and diabetes are on the rise, deaths from heart disease have been falling steadily since the 1950s – probably not coincidentally. The latest nutrition guidelines from the US Department of Health look very much like something Keys would have advocated. If you want to eat healthily, you can do a lot worse than follow his advice.

THE TRUTH ABOUT NUTRITION SCIENCE

At this point you can be forgiven for feeling a little overwhelmed and confused. I apologise if I've made everything seem less clear rather than more. But that is the unvarnished truth about nutrition science – it is very, very complicated. Of course, we know that what we eat affects our health. But discerning precisely how is almost impossible. Why this is so is worth dwelling on in detail.

Back in 2002, the WHO and the UN's Food and Agriculture Organization decided it was time to produce the definitive report on diet, nutrition and health. They convened an expert panel that reviewed more than 400 studies and wrote them up in an exhaustive report. The exercise was largely motivated by the fact that chronic diseases kill around 26 million people a year, and that bad diet is thought to be one of their main causes.

If the WHO and FAO were hoping for clear answers, they were to be sorely disappointed. Out of 140 possible links between nutrition and the 'big four' diet-related diseases – cancer, cardiovascular disease, osteoporosis and diabetes – only ten had enough evidence behind them to be deemed 'convincing'. All the report could say for sure was that eating too much fat and salt will increase your risk of cardiovascular disease while fruit, vegetables and oily fish reduce it; salted fish raises your risk of nasopharyngeal cancer; and if you're over fifty and want to avoid osteoporosis, you should increase your calcium and vitamin D intake. That's it. And even these 'conclusive' findings have been eroded since the report came out.

Why are there so few answers? One source of trouble is that nutritional research often focuses on single dietary factors – nutrients isolated from food, or specific food groups. This nutrient-by-nutrient analysis is valuable because it forms the basis of nutritional advice, such as the US Department of Agriculture's food pyramid. Yet any single nutrient isn't going to make a decisive difference to your health when you need at least fifty.

Also, people don't eat single nutrients or foodstuffs. They eat a mixed and varied diet consisting of many different foods, bundled into three-dimensional biological structures, which interact with each other in complex and unexpected ways. On top of that are lifestyle choices that also influence health and are difficult to separate from diet. You might find, for example, that people who eat a lot of fresh tropical fruit have fewer heart attacks. But fresh tropical fruit is expensive, so chances are that those who eat a lot of it are affluent and so can afford a gym membership or live further from a main road or take lots of relaxing holidays. Is it really the fruit that lowers heart attack risk, or is it a more general benefit of being wealthy?

These problems are amplified by the fact that humans are such difficult research subjects. There are basically three types of experiment you can do, each with some pros and many cons. The most stringent are metabolic studies, where researchers have complete control over participants' lives (including what they eat) for days or weeks at a time. These are useful for determining whether a certain nutrient, food or diet affects measurable biological factors, such as cholesterol level, that are believed to be linked with certain diseases. The catch is that these trials don't mimic real life and generally do not last long enough to provide practical knowledge about links between diet and health.

At the other end of the spectrum are observational studies, where researchers recruit a big group of healthy people and try to record what they eat 'in the wild' for months or years while monitoring their health. If at the end you find that people who ate a lot of peas, for example, have fewer cases of ingrowing toenails than people who eat no peas, you might reasonably conclude that peas prevent ingrowing toenails. (They don't, of course – this is just an illustrative example.)

Observational studies are the most common type, but are haunted by methodological gremlins. Monitoring what people eat is time-consuming so most studies rely on self-reporting, often in the form of a food diary. Unsurprisingly, these are highly unreliable. People forget to fill them in, deliberately or unconsciously omit lapses such as eating an entire packet of custard creams, or exaggerate their intake of healthy foods. There are also those confounding lifestyle factors – exercise, smoking, alcohol intake and myriad other variables – that statisticians must account for before they can work out what the results mean. But identifying and accounting for all confounders is all but impossible.

There are some standard corrections that can be applied. People in observational studies tend to under-report their calorie intake by about 25 per cent. You can account for that. But researchers rarely know what foods make up those unreported calories, and if you don't know what people are actually eating, it's hard to draw conclusions about their health.

The middle ground is the randomised intervention trial. In these studies one group is asked to change one aspect of their diet – eat less fat, say, or more fruit – for months or years on end, while the other group (called the control group) is told to carry on as normal. At the end, the researchers can compare the number of cases of, say, stomach cancer in the two groups to see whether the change in diet has had any effect. These trials are considered the most valid. But they are very expensive to carry out and hence rare.

And even though randomised intervention trials are the *most* valid, that doesn't make them *totally* valid. Such studies are hard to do well. The main problem is compliance: initially the experimental subjects are conscientious about following their dietary instructions, but as the study progresses they tend to backslide, while pretending that they are not (these trials often also rely on self-reporting). Control subjects, meanwhile, are free to change their diets voluntarily in response to health messages. The result is often a gradual convergence of the two groups' diets. And that creates a dilemma: to get statistically significant results, studies must extend the experiment for as long as possible, yet the longer the study goes on the more the difference between the two groups narrows. For this reason the perfect randomised trial is impossible to conduct.

According to one authoritative estimate, of the roughly one million papers that have been published in nutrition, just a few hundred are of the highest scientific quality. The rest are either small or poorly designed trials, opinion pieces, or reviews summarising the results of other research. Nutritional knowledge as a whole is overwhelmingly flawed, biased and confounded.

These methodological problems are compounded by yet another source of error known as publication bias. Studies producing unexpected and hence newsworthy results are more likely to get published in an academic journal than those that don't. Say there are two studies looking at the link between artificial sweeteners and obesity. One surprisingly finds that sweeteners cause obesity, the other that they prevent it. Which is more likely to be published? Meanwhile, a third study that finds a null result – that there is no link at all between sweeteners and obesity – is unlikely ever to be published at all.

These biases have a huge influence on what makes it into the mass media, and hence into the public consciousness. On this charge, universities and academic publishers are just as guilty as the media, as they are the ones seeking coverage by pumping out press releases to time-stressed journalists. The old adage about man bites dog (news) versus dog bites man (not news) is apposite here. One small and flawed study that goes against the prevailing wisdom will make headlines. A dozen large, scientifically sound ones that confirm the orthodoxy will sink without a trace.

This is the main reason why nutritional advice seems to flip-flop all the time, with, say, eggs or red wine or chocolate being good for you one week but bad the next. It is also the reason why nutrition researchers stress the need to look at the totality of the evidence rather than focusing on one, probably flawed, study.

So what does the totality of the evidence say? Unfortunately for people who want miracles or quick fixes, there are none. In fact, a few critics of nutrition research say the totality of evidence is so weak that there is little point in building your diet on it. Within certain common-sense boundaries, it doesn't really matter what you eat. The take-home message is that most of the orthodox advice – eating less fat, salt and sugar and more whole grains, fruit and vegetables – at least won't do you any harm, and will probably do you some good. This looks pretty similar to the Mediterranean diet, which we saw is still holding up quite well. It is also clear that the way to avoid obesity is to eat less and exercise more.

While I was putting the finishing touches to this book, the UK government's Scientific Advisory Committee on Nutrition released a long-awaited and exhaustive review of the evidence on saturated fats and health.[40] The work was commissioned in response to the research we reviewed at the start of this chapter, and took the best part of five years to complete.

The conclusion was unequivocal: saturated fats are bad for your health and should be kept below 10 per cent of your daily energy intake (the average in the UK is currently about 13 per cent). To reduce bad cholesterol levels and hence lower your risk of cardiovascular disease, go easy on the butter, cream, fatty meat and cheese. In other words, the advice on saturated fat does not need to change.

But the report did make an interesting recommendation – that getting down to 10 per cent saturated fats should be achieved by replacing them with unsaturated fats rather than carbohydrates, especially sugar. It thus acknowledges that our understanding of two major food groups – fats and carbs – has changed. Saturated fats can no longer be regarded as a health risk per se; what matters is the balance between saturated and unsaturated fat. The old dietary advice to replace saturated fat with starchy carbs should be thrown to the wolves.

So if the message you were hoping for from this chapter is 'eat fat', you've got it. But that does not mean steak, cream pies and hot fudge. It means the unsaturated fats in oily fish, vegetable oils, nuts and seeds.

One thing the report did not straighten out is whether saturated fats in dairy are somehow healthier than saturated fats in meat. That is one for the 'more research needed' files. For now, the advice is to go easy on all saturated fats, including those in whole milk, butter, cream and cheese.

Nutritionists queued up to comment about the findings and make wider dietary recommendations based on them. A well as cutting down on saturated fats and replacing them with unsaturated ones, the consensus was that a healthy diet should be low in added sugar and salt, high in fibre and rich in fruits, vegetables, whole grains, pulses and oily fish. In other words, a Mediterranean diet. As somebody from a Mediterranean country once said, *plus ça change, plus c'est la même chose.*

THE TRUTH
ABOUT DIETS
AND WEIGHT LOSS

TO MOST PEOPLE 'diet' is synonymous with losing weight. But there's more to dieting than cutting down on calories. A diet is any regime that restricts what you can and can't eat, usually for health reasons. Sometimes it is out of necessity – people with severe food allergies, for example, have no choice but to avoid certain foods. But more often than not people go on diets to achieve a positive health goal. And there is no shortage of options, from 5:2 fasting to veganism. It is easy to dismiss these as fads, but in some cases they genuinely work.

Of course, there are a lot of fad diets out there that have no basis in reality. The alkaline diet, for example, is based on the bizarre and unscientific claim that certain foods lower the pH of the blood – which is naturally slightly alkaline – and so should be avoided in order to prevent cancer and other diseases. This is total, absolute and utter nonsense. The same goes for dozens of other diets. The rule of this chapter is that if I don't mention a diet, it isn't even worth debunking.

But before we get into the nitty-gritty, it is helpful to know what constitutes a baseline adequate diet. This varies from person to person but as a rule of thumb you need to get enough calories to cover your energy expenditure (usually from carbo-hydrates), enough protein and fat to meet your body's structural and functional needs, plus various micronutrients such as vitamins and minerals.

Not everyone's calorie needs are the same: Michael Phelps

reportedly ate 8,000 calories a day while training to swim in the Olympics. For most of us mere mortals, however, the standard advice is that men should eat 2,500 calories a day and women 2,000, with 39 per cent of the calories coming from starchy carbohydrates such as bread, pasta, potatoes and rice (preferably wholegrain or high-fibre versions), 40 per cent from fruits and vegetables, 12 per cent from protein such as meat, fish, eggs and pulses, 8 per cent from milk and cheese, and just 1 per cent from added fats such as oils and spreads.

It's a simple recipe, but is surprisingly hard to stick to. Left to our own devices most of us tend to eat too many calories, too much of the tempting treats and not enough of the virtuous stuff. The result is weight gain, bad digestion and a less than optimal diet. Hence the popularity of periodically or permanently denying ourselves in increasingly elaborate ways, or forcing down things we don't really want to in the belief that they will do us good. But let's be frank: all of this takes willpower and can be tedious. So what, if any, forms of dietary restraint are worth the effort?

THE TRUTH ABOUT BREAKFAST

Let's start at the beginning. Breakfast is often said to be the most important meal of the day. Eating soon after you get up will make you livelier in mind and body, set you up to eat more healthily for the rest of the day, and ultimately make you slimmer; skip it at your peril.

On the other hand, some people swear off breakfast, claiming that this is the route to good health and slimness. Quite apart from the fact that what people eat for breakfast varies enormously – from a colossal full English to a piece of fruit – can it really be true that eating or skipping a single meal can have such major impacts?

The idea of specialised breakfast food is a relatively recent one, pioneered at the turn of the twentieth century by an American doctor called John Harvey Kellogg. At the time most people ate a normal cooked meal in the morning to fuel a hard day's labour. Kellogg invented the breakfast cereal cornflakes because he thought that eating fibre would promote good digestion and regular bowel movements, both health obsessions of the day. Legend has it that for some reason he also believed it would discourage masturbation. The marketing of other breakfast staples, including bacon, eggs, orange juice and coffee, soon followed.

Whatever you eat, the word breakfast refers to breaking what is usually the most prolonged period of fasting during a typical twenty-four-hour cycle. From what we know about fasting, by the time you sit down to your cereal/toast/fry-up, your blood sugar will be at rock bottom and you'll be making inroads into your energy savings, the glycogen stored in your muscles and liver. Glycogen is a carbohydrate made by polymerising glucose; it is like a current account for energy that can be dipped into quickly and easily.

It therefore makes sense that replenishing your energy will give you a physical boost. However, the claim about skipping breakfast making you fat seems paradoxical – how can missing a meal cause you to gain weight? This is explained away by the claim that people who skip breakfast overcompensate later in the day and end up eating more overall, and/or that people who do eat breakfast boost their metabolic rate and so burn more calories through the day (for more on metabolism, see page 118).

There's no shortage of research on these claims but most of it comes from observational studies, in which investigators watch people going about their normal life, without control groups. This means that other elements of their lifestyles – such as getting lots of exercise or sleep – could be the cause of the health effects. Which leaves a big unanswered question: are people healthy because they eat breakfast, or do they eat breakfast because they are healthy?

To separate the (shredded) wheat from the chaff and determine what, if any, causal effect breakfast can have on health, you need randomised controlled trials. In recent years scientists working on the Bath Breakfast Project – which sounds like a study on eating breakfast while having a bath but is actually just based at the University of Bath – have done such trials. In one, for example, they took two groups of lean adults and gave one of them breakfast while the other consumed nothing but water until midday.[1] Those in the breakfast club had to eat a big one – 700-calories or more. The scientists then took a range of measurements throughout the day, over the course of six weeks.

The results challenged almost everything we thought we knew. The fasters did not, in fact, overcompensate later in the day. They ate a bigger lunch but not enough to make up the 700-calorie deficit, and the breakfast group ended up eating more overall.

The scientists also measured ghrelin, a hunger hormone. At lunchtime they found that levels in both groups were much the same, which might explain why those who fasted didn't overeat. But after lunch the levels diverged. In those who had fasted it dropped, but it stayed high in those who had breakfasted – which calls into question the idea that breakfast sets you up to eat more healthily throughout the day. As for metabolic rate, eating breakfast did increase it a bit, but nowhere near enough to offset the 700 calories.

The breakfast eaters did do something unexpected and potentially beneficial. The team used accelerometers to measure their movements, especially small ones such as fidgeting, otherwise known as 'non-exercise activity thermogenesis', or NEAT. Over the course of a day this light activity can make a big difference to how much energy you use. The people who ate breakfast fidgeted more and burned significantly more calories from NEAT – several hundred more calories, in fact. The lack of fidgeting in the fasting group was enough to wipe out any gains from consuming fewer calories (for more on the benefits of fidgeting, see page 216). Unsurprisingly, then, over the six weeks there was no difference between the two groups in terms of weight gain or fat levels.

These findings have since been replicated in other studies; for example, using obese rather than lean subjects.[2] Overall, the project scientists conclude that 'current evidence does not support a clear effect of regularly consuming or skipping breakfast'.

But even if breakfast's reputation as the most important meal of the day has not been proven, we shouldn't throw away our toast racks and egg cups just yet. 'Breakfast' covers a multitude of food choices, and globally people eat a stunning variety of foods first thing. It may be that certain foods, say those rich in fats and protein, or alternatively carbs, really do produce the claimed benefits. Caffeine is also a centrepiece (or sole component) of many people's breakfasts; maybe it encourages fidgeting even in people who skip breakfast. As yet, we don't know.

In some ways the lack of clear evidence about breakfast is good news. If you eat it because you have swallowed the hype rather than actually wanting to swallow the food, try going without and see how you feel. That may open the door to trying intermittent fasting, which, while far from a sure-fire health boon, does appear to have benefits.

In any case, most of us could do with eating less, and given that breakfast is probably the easiest meal to skip, maybe mindlessly eating it every day is a missed opportunity.

THE TRUTH ABOUT FASTING

Which brings us neatly to the next diet. One of the leading food fads of recent years doesn't actually involve much food. In health-conscious circles there's growing buzz about intermittent fasting, such as the 5:2 diet where you eat normally for five days a week and a starvation diet for the other two; or the 16:8 where you fast completely for sixteen hours and eat anything you like during an eight-hour window, and alternate-day fasting where you eat as much as you like one day and fast the next.

If this all sounds like torture, bear in mind that the claimed pay-offs are enormous. Advocates of intermittent fasting – who include many of the scientists who study it, as well as those who research the biology of ageing – say these diets deliver many desirable benefits: weight loss; slower ageing; reduced risk of cancer, heart disease, diabetes, Parkinson's and Alzheimer's; a clearer head; and the general promise of a healthier life. But they demand sacrifices that are hard to stomach. So should you start fasting?

Intermittent fasting has its roots in a practice that has been known for decades to extend lifespan and improve metabolic health in animals. Caloric restriction (increasingly called dietary restriction, for reasons we'll get to later) involves a permanent reduction in food intake by up to 60 per cent. Hundreds of studies have found that organisms from single-celled yeasts to mammals are healthier, age more slowly and live longer when their calorie intake is capped. It works on every animal it has been tried on, including dogs and rhesus monkeys.

Caloric restriction has never really caught on in humans, however, because most people find it impossible to maintain for long periods. The best that has been achieved in experiments is a 10 per cent reduction in energy intake for a few months. However, a few hardy people do manage to maintain it voluntarily for years, and they show clear improvements in cardiovascular disease risk and other markers of health. It is tempting to joke that fasting doesn't make you live longer, it just feels like it does. But based on these admittedly small and unscientific self-experiments, it probably would extend your life, if you could keep it up.

That is where intermittent fasting comes in. It is essentially a tolerable way to achieve the benefits of caloric restriction without demanding permanent self-restraint.

Fasting regimes

There are numerous regimes available. The easiest and most flexible is the 16:8. Sleep time is included so, for example, you can start fasting at 8 p.m. and stop at midday the next day. Think of it simply as missing breakfast. You can do this as frequently as you please, but the more you do it the better the results – say advocates.

On fast days of the 5:2 diet you are allowed a single 600-calorie meal, so a bit of calorie counting is required. A more spartan regimen has similar restricted calorie fasts every other day. Then there's total fasting, where you consume no calories for anything from twenty-four hours to five days (longer than about a week is considered potentially dangerous). All of these regimes can be a one-off, or be repeated weekly or monthly.

During fasting you are allowed to drink water and other zero-calorie drinks such as black tea and coffee (no sugar, no milk), and diet sodas – make sure they really are zero calorie rather than just low calorie. Alcohol is strictly forbidden.

A more forgiving regime is the fasting-mimicking diet, which claims to deliver the benefits of longer-term fasting without abstinence from food. You purchase a diet plan that keeps you on a strict regime for five days each month – mostly soup, plus a few crackers, olives and the odd nut bar. It is essentially intermittent caloric restriction.

The goal of all these diets is to put your body into a metabolic state that promotes cellular repair and waste disposal. Growing evidence seems to support that periodically going without food puts our bodies into a kind of emergency mode, where we conserve energy, make repairs and prioritise physical and mental functions that maximise the chances of finding food.

There's an evolutionary rationale to this. We may live in a technologically advanced civilisation but our physiology is still stuck in the Palaeolithic, when our ancestors were opportunistic hunter-gatherers and probably had to endure frequent bouts of famine, punctuated by occasional feasts. By this reckoning, fasting from time to time is a more natural state than eating three square meals a day.

Thanks to animal studies, we know what happens in the body when fasting kicks in. At first you burn the glucose in your bloodstream, but that does not last long. Then you start on the glycogen stored in your liver. Once that is depleted, a process called autophagy starts, where damaged cells, organelles and proteins are broken down for fuel. This system probably evolved to maximise the chance of surviving famine, but also cleans out cellular trash that would otherwise build up and cause problems.

Autophagy becomes less efficient as we age, which causes detritus to build up inside cells, a process that has been linked to many age-related diseases including cancer, and to the ageing process itself.

You don't have to deprive yourself for long to get autophagy working. The liver stores about 700 calories' worth of glycogen and routine daily activity burns around seventy calories an hour, which means that after about ten hours your glycogen is all gone and you're in repair mode. On the 16:8 fast, that gives you a full six hours of autophagy.

Even when you are asleep, you are burning about sixty calories an hour, and also technically fasting. Add exercise into the equation and the switch can happen even faster. A vigorous thirty-minute run can burn 300 calories, which means you're almost halfway to autophagy.

Further down the road another emergency system kicks in. The liver starts converting stored fats into 'ketone bodies' – short molecules that are by-products of the breakdown of fatty acids, which can be used by the brain as fuel. This process, called ketosis, is in full swing three to four days into a fast, so is not generally triggered by intermittent fasting. It is of dubious health benefit, though some people eat a diet deliberately designed to trigger it. That means getting about 70 per cent of calories from fat and most of the rest from protein, very like the Atkins diet. There is some evidence that it can work for weight loss but whether that is down to ketosis itself or just to eating less is not clear.

Various hormones are also affected by intermittent fasting. Production of both insulin and IGF-1 (insulin-like growth factor 1) drops. High levels of both have been linked to cancer, and elevated insulin is also associated with type 2 diabetes. Fasting also tends to mean eating a lot less animal protein and fat, which have both been linked to cancer.

Many people are surprised at how easy intermittent fasting is, and report feeling energised, wired and alert during and after a fast. But there can be downsides. Researchers accept that going without food can leave some people feeling rotten in the short term, with complaints such as fuzzy-headedness, tiredness, blurred vision and stomach-ache.

So is it worth it? As yet there is scant data from large-scale human trials, and what there is does not support the grandiose claims of the fasting fanatics. But for people who could benefit from dropping a few pounds, some forms of intermittent fasting look like a good option. A study of obese men found that sticking to a 5:2 diet over six months produced the same weight loss as a more conventional calorie-controlled diet demanding 24/7 abstemiousness.[3] Fasting also led to greater improvements in blood-sugar control.

A different human trial of a fasting-mimicking diet found that people who stuck to it for three cycles dropped body weight and visceral fat, and ended up with healthier blood pressure and lower levels of IGF-1, cholesterol and C-reactive protein (CRP), a marker of inflammation.[4] In this trial, two-thirds of the participants were overweight or obese.

For people who are overweight, then, any kind of intermittent fasting diet will probably help weight loss and hence cut the risk of diabetes and cardiovascular problems. For people who are healthy to begin with, though, does fasting offer a benefit beyond the fact you inevitably cut a few calories and so may lose a bit of weight? Based on current research, we just don't know.

Mouse experiments, however, support the idea that bouts of fasting are beneficial. If non-obese mice are put on a permanent 16:8 diet they end up leaner and fitter than mice allowed to eat freely 24/7.[5] Weirdly, the time-restricted mice ate a little bit more overall. Mice on the 16:8 regime during the week but given weekends off saw similar gains, which is encouraging news for people who like to let their hair down. But mice are not humans, and humans don't eat laboratory mouse chow, so how relevant these findings really are to us is not known. Intriguingly, research on the effect of humans routinely skipping breakfast – which is effectively doing a 16:8 – has found no effect on weight loss or body fat.

Another claim about fasting is that it boosts cognitive performance. The brain is not the topic of this book but this claim is worth dwelling on a bit. Fasters regularly say they achieve clarity of thought and sharper focus. This makes sense from an evolutionary perspective, because if you are deprived of food you need a clear head to figure out how to find some. It also fits with observations of hunter-gatherers. The Wopkaimin of Papua New Guinea, for example, prefer to hunt on an empty stomach because they say it makes them sharper. So far, though, no controlled studies have been done to investigate the link between fasting and cognition in humans.

One potentially tasty possibility is that the benefits of caloric restriction are available without actually restricting calories. Some early-stage science suggests that cutting calories per se is not what produces the goods. Instead, their reduction is simply a by-product of cutting certain nutrients as a result of the diet.

That is why the term 'caloric restriction' is being superseded by 'dietary restriction'.

The nutrients cut are a group of amino acids – the building blocks of proteins – called the branched-chain amino acids. There are three of them: leucine, isoleucine and valine. For unknown reasons mice fed a diet restricted in these are leaner and fitter, even if they eat the same number of calories as mice not on the diet.

This is in keeping with epidemiological evidence that people who eat a low-protein diet have lower rates of cancer, diabetes and overall mortality. And in a randomised human trial of a protein-restricted diet, people who obtained fewer than 10 per cent of their calories from protein over six weeks lost about 2.5 kilograms and had improved metabolic health, including a significant decrease in fasting blood glucose, compared to people who ate a typical Western diet with 17 per cent of calories from protein.[6] The low-protein group actually ate about 10 per cent more overall.

That suggests that a low-protein diet is beneficial, though don't try it if you are over the age of sixty-five. At that point loss of muscle mass is a bigger health risk than too much protein in the diet, so for seniors a high-protein diet is required to maintain muscle mass.

A low-protein diet goes against received wisdom that protein is highly satiating and so can help you lose weight by reducing overall food intake. But there may be a way to square this circle. It should be possible to avoid foods high in those three amino acids without avoiding protein altogether. Turkey meat, for example, has low levels of leucine, isoleucine and valine.

THE TRUTH ABOUT DETOX

Another reason that people sometimes give for fasting – or embarking on some other restrictive diet – is that they are 'detoxing'.

Few ideas in health and well-being are as popular as the notion that our bodies need a regular detox. You can see why: it is intuitively appealing, because we are under siege from toxins. The air we breathe, the water we drink and the food we eat are loaded with pollutants. We eat junk food, wash it down with booze, guzzle additives and caffeine, take pharmaceutical and recreational drugs, and douse ourselves in shampoo, shower gel and cosmetics. Swamped by these synthetic chemicals, our natural detoxification systems – the liver, kidneys and gut – cannot cope. Once the system becomes overloaded, toxins build up in our bodies and trouble follows. The idea is so entrenched that some pubs try to pull in punters with signs saying 'Retox here!'

It's all bunkum. The boring truth is that your natural detox systems are perfectly capable of dealing with the vast majority of toxins that modern life throws at you. As long as your liver, kidneys and gut are working properly most of the toxic chemicals we consume are broken down and excreted within hours. There is no scientific evidence that detox diets rid us of anything other than money.[7] Some may be positively harmful, exacerbating the problems they claim to solve.

It is true that some pollutants, many of them toxic, do accumulate in our bodies. These so-called persistent organic pollutants (POPs) include pesticides, dioxins and polychlorinated biphenyls (PCBs). If we take these in faster than our bodies can get rid of them, levels build up in our bodies – especially in fat. But detoxing won't help. To get rid of something like PCBs, only zero exposure for prolonged periods will work. Even then it would take the best part of a decade to reduce the quantity in your tissues by half. And that is impossible because there is no way we can reduce our exposure to zero.

What's more, extreme fasting or dieting – a central plank of many detox programmes – can release fat-soluble chemicals into the blood rather than eliminating them from the body. One study found the level of pesticides in blood shot up by 25 to 50 per cent after people dropped a lot of weight quickly.[8] Animal studies show that this increases the concentration of compounds in tissues like the muscles and brain, where they can do more harm than when safely sequestered in fat. This sudden flood of chemicals could create the exact problems that detoxers are trying to avoid. There's also no guarantee that chemicals released from fat will actually leave the body – some will end up back in storage.

The NHS warns that detox could be harmful. Many herbal supplements contain ingredients that have never been put through human safety trials and could interfere with medicines. There's also the danger that detox gives people a licence to live unhealthily, comforted by the mistaken belief that they can atone later.

Of course, it won't do you any harm to drink less, stop smoking, drink plenty of water and eat more healthily. As we have already seen, periods of intermittent fasting appear to have benefits too. But if you are still tempted by the siren call of a detox diet, consider the wise words of the British Dietetic Association: 'the idea of "detox" is a load of nonsense . . . for the vast majority of people, a sensible diet and regular physical activity really are the only ways to properly protect your health'.

THE TRUTH ABOUT PALAEO

Going 'palaeo' is another diet craze that needs a reality check. The idea that reverting to our ancestral diet – eating only the things that were on the menu in Africa 50,000 years ago – is a sure-fire route to health is both plausible and appealing. After all, surely that is what our bodies evolved to survive and thrive on. But what constitutes a palaeo diet, and is it really all it is cracked up to be?

The palaeo diet takes its name from the Palaeolithic era (or Old Stone Age). It has its roots in the 'evolutionary discordance hypothesis', put forward in 1985 by a medic and anthropologist at Emory University in Atlanta, Georgia.[9] They pointed out that while our genes have barely changed for at least 50,000 years, our diets and lifestyles were totally transformed by the invention of agriculture 12,000 years ago or so – too quickly for evolution to have adapted us to it. This, they said, is the reason why obesity, diabetes, heart disease and cancers are rife. If we could only eat like hunter-gatherers on the East African savannah before the agricultural revolution, we'd be healthier.

In recent years, going palaeo has become very popular. It involves giving up grains, milk, legumes (yes, lentils are a modern curse), oils, refined carbs, alcohol, caffeine and added salt, and eating only meat, offal, fish, seafood, eggs, fruit, vegetables, roots and nuts, often raw.

Some of this, such as eating smaller amounts of highly processed grains and sugars, is sound nutritional advice. But a lot is not. Whole grains, legumes and milk can definitely be part of a healthy diet. And in any case the underlying rationale is wishful thinking and bad science.

The idea that humans were living in perfect harmony with their diets 50,000 years ago and are now out of whack is romantic tosh. Environments change all the time so it is unlikely that hunter-gatherers were perfectly evolved for anything. We, meanwhile, have adapted somewhat to our agricultural diet. Many people now have extra copies of genes for digesting the starch found in grains, and the ability to digest milk as an adult evolved independently in several populations.

In any case we can't know for sure what our ancestors ate, and in what quantities. Did they eat meat all the time, or rarely? Was fish a major component of their diets? The archaeological record shows that there was no such thing as 'The' palaeo diet. Our ancestors ate a much wider range and variety of foods than you will find in a palaeo cookbook – including cereals, which were eaten by our ancestors many millennia before they were domesticated.

What we can say for sure is that they didn't eat anything like the animals and plants we eat today, which mostly did not originate in Africa and which have been transformed beyond recognition by selective breeding and modern farming methods. So even if you stick to the palaeo diet, you're not eating an actual palaeo diet, unless you are foraging and hunting it from the savannahs of East Africa.

Last but not least, it's not clear that our ancestors were free of modern diet-related diseases. An analysis of five mummies from the Unangan culture of the Aleutian Islands in the Bering Strait found that three of them had atherosclerosis.[10] The Unangans were hunter-gatherers living mostly on seafood. Oysters, anyone?

On the plus side, the palaeo diet is clearly better than eating lots of junk food, and steering clear of alcohol is a good idea. It is also one of many that have been found to be effective for weight loss, but only when it is hypocaloric – in other words, leads to you eating fewer calories than you use. One thing we can safely surmise about our hunter-gatherer ancestors is that they were leaner on average than we are. But they were probably also hungry quite a lot of the time.

THE TRUTH ABOUT GOING VEGAN

We can probably guess that one diet our ancestors did follow from time to time – largely out of necessity – was one containing no animal products. Nowadays a growing number of people do this by choice, usually for ethical and/or environmental reasons but also because they think it is healthier to give up meat, fish and dairy.

When the British bakery chain Greggs launched a vegan sausage roll in 2019 and saw its profits spike as a result, something had clearly changed. Once the preserve of a hardy few, the butt of jokes and insults, veganism has now exploded into the mainstream. Celebrities from Natalie Portman and Serena Williams to Lewis Hamilton have all come out as vegan. Supermarkets and restaurants are scrambling onto the plant-power bandwagon.

In 2014 vegans made up just 1 per cent of the US population. Three years on, an additional sixteen million – 5 per cent of the nation – had joined the club. In the UK, numbers are smaller but also growing. A 2016 poll suggests that just over 1 per cent of Britons never eat meat or animal products. According to the UK Vegan Society, that's a more than threefold increase in ten years. For many, the main motivation is animal welfare and the environment. There is no doubt that a vegan diet is better for both. But it is also a healthier way to live, if you follow a few simple precautions.

Vegans eat a lot of fibre, vitamins, minerals and antioxidants, and not a lot of saturated fat or cholesterol. Even compared with vegetarians, vegans tend to be leaner, with lower blood pressure, healthier cholesterol levels and a reduced risk of cardio-vascular disease.[11]

But following a vegan diet can be a trial. Strict veganism means giving up all products derived from animals. That obviously means no meat, fish, dairy or eggs – which translates into avoiding huge numbers of ready-made products. Most vegans also avoid honey, because bees produce it, and wine, because animal by-products such as gelatin and isinglass (derived from fish) are used as fining agents. Leather, wool and silk are also off limits; some vegans refuse all products tested on animals. But let's stick to food and drink.

For people considering going vegan, the dietary constraints can raise health concerns. Meat, fish, dairy and eggs supply nutrients that can be difficult to obtain elsewhere. The clichéd view of vegans as weaklings chewing joylessly on lentils is offensive nonsense, but captures a fear that many have about quitting animal products. Humans evolved on an omnivorous diet, so can we really get everything we need – including pleasure – from plants?

The short answer is yes, but you will need discipline to tick all the nutritional boxes and ingenuity to make the diet interesting.

Macronutrients are easiest. Carbohydrates are the least of your worries, as most of them come from plant sources anyway: grains, starchy vegetables, legumes and fruit. Fats are fairly easy too – think vegetable oils, nuts and avocados. Protein is abundant in pulses, seeds, quinoa and tofu.

However, covering all the bases still requires work. Cutting out all animal products may be healthier overall but it does increase the risk of certain nutritional deficiencies. Human biochemistry is versatile but there are some nutrients that it cannot rustle up from other molecules and which must therefore be included in the diet – what are known as 'essential' nutrients. Two of these are fatty acids – alpha-linolenic acid (an omega-3 fatty acid) and linolenic acid (an omega-6 fatty acid) – that serve multiple vital bodily functions. In an omnivorous diet, fish and seafood mainly supply them. But they are also found in many plant sources, notably leafy vegetables, seaweed, olive oil, soya oil, seeds and their oils (especially linseed/flax, pumpkin, sunflower, hemp, chia and rape), and walnuts.

Two other omega-3 fatty acids are borderline essential. Eicosapentaenoic acid (EPA) and docosahexaenoic acid (DHA) can both be synthesised from ALA but only at low efficiencies that may not supply enough, requiring dietary sources. Unfortunately for vegans, by far the best source is fish. Many foods are fortified with them, though often with fish oils. However, omega-3/6 food supplements made from algae are also available.

There are also nine essential amino acids (the building blocks of proteins) that have to be included in your diet. Meat, fish and dairy are chock full of them but again plants can supply more than enough. As long as you're eating plenty of pulses, nuts and seeds you should be fine; tofu and other soy-based meat and dairy substitutes are also a good source.

More problematic are micronutrients. Top of this list is vitamin B12, a deficiency of which can cause anaemia and nerve damage. B12 is made by bacteria, and omnivores get it from meat, fish, milk and eggs. Plants do not make it at all, and the only fail-safe sources for vegans are fortified foods and supplements. B12 deficiency is also a risk for vegetarians (for more on dietary supplements, see page 125).

Fortified breakfast cereals are a good way to fill the gap; many non-dairy substitutes are also fortified. Nutritional yeast flakes are another good source; they look unappetising but actually have a pleasantly umami flavour that some compare to Parmesan. Supplements are also available. You don't need a lot to avoid deficiency, but if you're thinking about going vegan you should seek expert advice. The NHS advises speaking to your doctor about getting all the nutrients you need.

Vitamin D is another worry, but that is also true for non-vegans. The best source for all of us is sunlight, but during the darker months in the UK there usually isn't enough for our skin to manufacture the recommended daily dose. Topping up from food is difficult. Dietary supplements are the answer, though some are derived from animal sources.

The final big issue is calcium. Without dairy, vegans have to rely on green vegetables, okra, figs, chia seeds and almonds. Calcium supplements are available but have recently been found to increase the risk of kidney stones, gastrointestinal problems and heart attacks, so are best avoided. Many foods are fortified with calcium, and for some reason fortification appears not to bring the same risk as supplements.

Vegans also need to keep an eye on their iodine and iron intake. Iodine, important for thyroid function and metabolism, can come from seaweed or cranberries. Iron is available in green vegetables, legumes, nuts and seeds. However, these contain a form that is harder for our bodies to use, so the US National Institutes of Health recommends that people who don't eat meat consume nearly twice as much diet-based iron as those who do.

THE TRUTH ABOUT DIETS AND WEIGHT LOSS

Children may need to be particularly careful of their diet. There have been reports of severe nutritional deficiencies and neurological and physiological disorders in children who are raised vegan.

Beyond these, a vegan diet should supply everything necessary. One upshot of the growth in veganism is that a new generation of plant-based substitutes for fast food staples such as burgers, fried chicken and hot dogs is now readily available. And with supermarkets and restaurants increasingly catering for vegans, and innovative and increasingly realistic animal-free substitutes such as 'bleeding' burgers, vegan cheeses and cakes coming onto the market, there's no need to see it as a pleasure-free zone. One thing to bear in mind, however, is that these meat and dairy alternatives are not always necessarily healthy – particularly those that are so heavily processed.

For example, a typical three-ounce plant-based burger contains around 220 calories, thirteen grams of fat and just over a gram of salt. In terms of nutritional sins, that is no better than a typical beef patty. In fact, while there is no cholesterol, it is actually heavier on the salt and saturated fat.

Nonetheless, when going vegan is done sensibly, the upsides – getting thinner and healthier, with a clearer conscience to boot – seem to be worth the sacrifice.

THE TRUTH ABOUT FOOD INTOLERANCE AND AVOIDANCE

There's a well-known phrase in well-being circles: the road to health is paved with good intestines. For millions of people, however, that is a pipe dream. Between 5 and 15 per cent of Westerners have irritable bowel syndrome (IBS), a catch-all term for a poorly understood constellation of gut symptoms including bloating, abdominal pain and attacks of diarrhoea or constipation. Many more people suffer from milder chronic gut disturbances. So what's causing them? And can restricting your diet help?

The wheat protein gluten has copped much of the blame, even though the vast majority of people can digest it with no problem. But other foods are also routinely blamed for bad guts. One is the milk sugar lactose, which about 5 per cent of people of northern European descent cannot digest. Other foods that people commonly report bowel problems after eating are cabbage, onions, peas, beans, tomatoes, cucumber, fruit of all kinds, cheese, processed meats, wine, fats, fried foods, herbs and hot spices. Caffeine is also known to be a trigger. All told, about 20 per cent of people report a food intolerance of one kind or another.

Medical professionals are often intolerant of intolerances because they sometimes see them as self-diagnosed symptoms of the worried well. Now, however, a better understanding is leading to greater acceptance.

Gluten and FODMAPs
Gluten remains the biggie. We dealt with the nutrition science of gluten on page 52, so now let's tackle intolerance. Is it plausible that something that has been a staple food for centuries should suddenly turn out to be so bad for so many? If so, should you consider jumping on the gluten-free bandwagon?

It is true that some people are violently intolerant of gluten. Most have a condition called coeliac disease in which the protein triggers an autoimmune reaction that produces antibodies. These attack the lining of the gut, leading to inflammation, diarrhoea, constipation, cramps, malnutrition and fatigue. Coeliac disease affects around 1 per cent of people, although many of them have only mild symptoms and so are often unaware of their condition.

Far rarer are wheat allergies, which involve an immune response to other proteins in wheat. Symptoms appear rapidly if wheat is eaten or flour dust breathed in, or sometimes if people exercise soon after eating wheat.

If you have health problems that you suspect are linked to wheat, the first thing to do is to get tested for coeliac disease and wheat allergy.

However, many people who test negative for these still complain that wheat products cause abdominal bloating, gut pain, headaches, fuzzy-headedness and lethargy. The common name for this is 'gluten intolerance', though its official name is non-coeliac gluten sensitivity (NCGS). There are claims that up to a fifth of people have it. However, its existence remains disputed.

Several small studies have found that some individuals do have gut-related symptoms when they eat wheat that often clear up if they go gluten free. That may seem like strong support for the existence of NCGS, but it doesn't rule out the possibility that something other than gluten is responsible. There is growing evidence that this is indeed the case. That something may be a group of carbohydrates collectively called FODMAPs – we'll get to these shortly.

The science is far from settled, however. There may be a subset of people who don't have coeliac disease but who do genuinely suffer from gut problems when they eat gluten, through an as yet undefined mechanism. If you want to try going gluten free, bear in mind that it means excluding a vast number of foods containing grains, which includes beer and huge numbers of processed foods to which gluten is added as a stabiliser.

Ketchup and ice cream, for example, often list gluten as an ingredient. There's also the fact that cutting out gluten totally is complicated, inconvenient and often expensive.

Alternatively, the cause might not be gluten but those afore-mentioned FODMAPs – carbohydrates that the small intestine finds hard to absorb, and so which hang about in the gut being fermented by bacteria. This creates copious amounts of gas, causing discomfort and, not to put too fine a point on it, farting.

Wheat is high in certain FODMAPs but so are a frighteningly large number of other foods and drinks, including onions, asparagus, peppers, apples, dried fruits, peas, honey, milk, ice cream, many sweeteners and beer. The worst offenders are onions, garlic, rye and barley – and wheat.

FODMAP is an acronym for fermentable oligosaccharides, disaccharides, monosaccharides and polyols. Most are sugar molecules, either singly or in chains. The oligosaccharide bit is mostly short polymers of fructose (fructans) or galactose (galacto-oligosaccharides). The disaccharides are mostly lactose; the monosaccharides fructose. Polyols are also known as sugar alcohols and include sorbitol, mannitol, xylitol and maltitol. All are used as low-calorie sweeteners.

So people who claim to be to wheat intolerant are not necessarily wrong: they may be sensitive not to gluten but to FODMAPs, especially the fructans found in wheat. Going gluten free inadvertently reduces exposure to FODMAPs by about 50 per cent.

That was illustrated by a recent trial in which adults who were following a gluten-free diet but did not have coeliac disease were fed cereal bars packed with either gluten (but no FODMAPs), fructans or neither.[12] Those who had the fructan bar reported more bloating and other gut symptoms compared with the other two groups. The gluten and control group were indistinguishable.

For up to 70 per cent of people with IBS, reducing intake of these foods provides relief from pain and bloating. The low-FODMAP diet is considered so effective for IBS that it is now recommended by the National Health Service in the UK and bodies including the Gastroenterological Society of Australia. The diet may also relieve other gastrointestinal conditions, including acid reflux and indigestion. However, going FODMAP free is not straightforward. They are found in so many foods that giving them up entirely is extremely challenging. Imagine having to give up not just wheat and other grains but also lots of fruits, vegetables, pulses and dairy.

Thankfully, the type and quantity of FODMAPs that trigger symptoms vary enormously between individuals, so many FODMAP-sensitive people only have to give up a few foods to gain the benefits. But working out what they are can be difficult. For some people it is as simple as quitting onions, which are very high in fructans. Others cut out all FODMAPs for a while, then gradually reintroduce foods one by one to see which they can tolerate. But many people struggle to work out which foods to eliminate. There are books and apps available to help design a low-FODMAP diet, but it is a complicated and challenging process, and generally needs the assistance of a professional dietician.

Although it is appealing to think that we might all benefit from fewer FODMAPs in our life, experts say that if you don't have bowel problems there is no rationale for being on a low-FODMAP diet, because it is not necessarily good for your health. One drawback is that it cuts out foods that are important sources of vitamins, fibre and nutrients. Recently, there have also been concerns that it depresses levels of good bacteria in the gut. The long-term consequences of cutting out FODMAPs remain unclear.

About a third of people with IBS who try the diet reap no benefits. For them, some researchers are now looking at a more controversial alternative. When it comes to gut health, one of the most common pieces of advice has been to eat plenty of fibre, such as wholegrain bread and fibrous vegetables, which help keep the gut working properly and can also reduce the risk of cancer and cardiovascular disease. But fibre may also be part of the problem. Some types of fibre release gas when fermented in the bowel, causing irritation in sensitive individuals.

Some early research demonstrates that reducing fibre could be beneficial in IBS, with a study finding that wheat bran made 55 per cent of people with IBS feel worse and only 10 per cent feel better.[13] The trouble is that advising patients to eliminate both FODMAPs and fibre flies in the face of good nutrition.

The good news for those with gut problems is that the diversity of symptoms and causes is being investigated like never before. For those who don't, however, it might be time to step off the bandwagon. If you don't have any symptoms, you don't need to be on any of these diets.

Lactose intolerance

Around 7,500 years ago, somewhere in central Europe, a random but advantageous genetic mutation began to spread through the population. The mutation was in a gene that codes for a digestive enzyme called lactase, which breaks down milk sugar (lactose). Normally the gene is permanently shut off in late childhood. The mutation kept it permanently on. For the first time in history, adult humans could digest milk.

It's easy to see how the milk-digesting mutation would have been beneficial, as milk is a rich source of fats, proteins and other nutrients, and would have been a valuable and possibly life-saving supplement to the diet.

Even so, the majority of adult humans worldwide still don't produce lactase and cannot digest milk. In people who lack lactase, lactose passes into the colon, where it feeds bacteria that generate gas and fluid, resulting in painful bloating, cramps and diarrhoea – a condition known as lactose intolerance or malabsorption. In China and South East Asia, more than 90 per cent of people are lactose intolerant, compared with between 2 and 20 per cent of those of northern European descent. However, its incidence is hard to pin down as many people self-diagnose rather than taking a clinical test.

Numerous home test kits claim to detect lactose intolerance, but not all are reliable or based on solid science. If you suspect you are intolerant, the best advice is to get your doctor to perform a clinical test – usually a breath test that detects the fermentation of lactose by gut bacteria.

If you are diagnosed, it may only be temporary; symptoms of lactose intolerance peak between the ages of ten and sixteen. You can also become temporarily lactose intolerant as a result of gastroenteritis, bowel injury and stress.

There are other misconceptions. For one thing, people who genuinely can't absorb lactose can still drink moderate amounts of milk without ill effects – up to 240 millilitres in a single sitting, perhaps twice this amount if spread throughout the day. They can also usually consume yoghurt and hard cheese as most of the lactose is broken down during production. Not so for goat's, buffalo's, sheep's or yak's milk, all of which contain similar levels of lactose to cow's milk. Finally, although some celebrities love to blame lactose intolerance for skin complaints, weight gain and asthma, none of these is likely to result from failing to absorb lactose in the gut.

If you are lactose intolerant and looking for alternatives to dairy, see page 50 for more detail on alt-milk substitutes.

THE TRUTH ABOUT PROBIOTICS

The quest for good intestines doesn't stop at FODMAPs; it features regular doses of what have come to be known (thanks to successful advertising) as 'good bacteria'.

Probiotics – foods prepared with live 'good' bacteria, such as yoghurt or supplements – have been shown to help with conditions like irritable bowel syndrome; they may also be beneficial for people with weakened immune systems. But their long-term effects on overall health are not yet known.

Fermented foods and drinks such as kimchi, sauerkraut, kombucha and kefir are gaining in popularity. Their effects on health again aren't yet clear, but consuming them won't do any harm and might improve the microbial diversity in your gut.

For people in good general health, prebiotics, which encourage the growth of good bacteria, are the way to go. Getting new bacteria to colonise your gut is difficult but encouraging the growth of existing beneficial ones somewhat easier.

The simplest way to feed a healthy gut is to eat a diversity of fruits and vegetables. But for prebiotics, good sources include green bananas, asparagus, vegetables in the sunflower family such as artichokes, radicchio, lettuce, chicory, tarragon and salsify; and alliums such as onions, leeks, garlic, shallots and chives. Just watch out for the FODMAPS.

THE TRUTH ABOUT LOSING WEIGHT (AND KEEPING IT OFF)

If you want to lose weight, the world is your oyster. Or maybe oysters are your world. According to any number of websites dishing out dietary advice, eating oysters is a great way of shedding the pounds.

Problem is, that is just one of thousands of pieces of advice out there. Other foods touted for weight loss include eggs, avocados, chillies, full-fat yoghurt, coconut oil, grapefruit, cider vinegar, all kinds of seeds and even boiled potatoes. There is no shortage of diet plans either. Atkins, Zone, Dukan, fasting, South Beach, bulletproof, vegan, ultra-low-fat, high-protein, raw food, ketogenic, palaeo . . . you pays your money and you takes your pick.

The scientific literature is similarly overwhelming, with literally thousands of studies looking at every conceivable weight-loss and weight-maintenance diet and often coming to inconclusive or contradictory conclusions. If you want to lose weight, how are you supposed to navigate your way through this information minefield?

There's no doubt that staying lean is good for your health. Being overweight or obese is associated with all sorts of negative health outcomes, from heart disease and diabetes to cancer, and the fatter you get, the higher the risks. Although it is possible to be fat and aerobically fit, this does not mean that being fat is not unhealthy. On average obese people are much more likely to develop cardiovascular diseases even if they don't have warning signs such as high blood pressure.

If you are overweight or obese (see definitions on page 111) it is a good idea to try to get your BMI down into the 'normal' range, and keep it there.

Hence the enduring popularity of weight-loss diets. But achieving – and maintaining – a healthy BMI is not straight-forward. We live in what nutrition scientists call an 'obesogenic' environment, with easy and cheap access to high-calorie food and drink, and ample opportunity to be idle. Global rates of obesity and overweight have doubled since 1980 and a third of the world is now classified as overweight or obese.

So here's the good news: faddy-sounding diets and quick fixes may actually work. In 2013 the American College of Cardiology and American Heart Association set up a joint task force to review all the evidence on weight-loss diets.[14] They looked at all kinds of diets and all kinds of studies – observational studies, randomised trials and meta-analyses – and concluded that diets succeed as long as they achieve one crucial factor: a negative energy balance.

A recent update of the 2013 review came to the same conclusion.[15] No weight-loss regime can definitively be said to be superior to any other. On average they all work equally well as long as they are 'hypocaloric', meaning they lead you to consume fewer calories than you use. Boring, but true. In order to lose weight, you can do pretty much anything as long as you eat (and drink) less and/or exercise more (though the relationship between exercise and weight loss is not as straightforward as it looks; for more on this, see page 213). How you do it is up to you, but reconcile yourself to the fact that it will probably entail some hunger.

As for maintaining a healthy weight, that is a slightly different proposition. As the fat comes off, moving your slimmer body requires less energy and your metabolism consequently slows down. That makes it increasingly hard to shed the last few pounds – the dietary law of diminishing returns – and can contribute to rebound when you stop dieting. This is because metabolism remains sluggish even if weight starts going back on. Most people who successfully lose weight regain all or most of it in a few months or years, unless they stick to a regime that balances energy in and energy out.

It is as if the body is fighting to return to the pre-weight-loss state, and it is. There is a good evolutionary reason for this. We have evolved to store enough energy to get through lean periods. So if food intake drops, your body assumes that there's a famine and attempts to hang on to its precious resources by slowing down metabolism, then keeps it low when good times return to ensure your larder – or lard – is restocked for the next famine.

THE TRUTH ABOUT BMI

If you want to lose weight, you need to weigh yourself. But how do you know what magic number to aim for? The easiest – though perhaps not the most accurate – tool is BMI. It is a crude but still useful measure of whether you are a healthy weight.

To calculate your BMI, divide your weight in kilograms by the square of your height in metres. If the answer is between 18.5 and 25, your weight is normal. If it is below 18.5 you're underweight; 25 to 30 overweight, and 30 plus obese.

BMI is often criticised because it fails to capture some important nuances of body shape and composition. For example, it doesn't estimate percentage body fat, or how much super-unhealthy abdominal fat you have. It consistently gives tall people a higher BMI than shorter people with the exact same body shape and composition. It can be different for men and women with the same level of body fat. BMI can be higher than normal even if you don't have much fat, especially if you are male and very muscular.

For this reason, various other measures have been proposed, such as the Body Adiposity Index, which directly estimates your percentage body fat. To calculate it measure your hip circumference in centimetres and divide it by your height in cm raised to the power 1.5, then subtract 18. The average for men is about 18–24, and for women 25–31. However, defining an ideal amount of body fat is not easy.

Because abdominal fat is especially risky for cardiovascular disease and type 2 diabetes, the NHS recommends also measuring your waist. To do so, find the midpoint between the bottom of your ribs and the top of your hips. Breathe out and measure the circumference. Regardless of height or BMI this should be no more than 94 centimetres (37 inches) for men and 80 centimetres (31.5 inches) for women.

Fat and fit?

Being very obese is extremely bad for you. It is called morbid obesity for good reason. Having a BMI of over 40 increases the risk of dying from any cause by up to 29 per cent. It also increases the risk of type 2 diabetes, heart disease and certain cancers.

But, weirdly, carrying just a few extra pounds seems to have the opposite effect. A recent review of nearly a hundred studies involving nearly three million people found that being 'overweight' – defined as having a BMI of 25 to 29 – delivers a 6 per cent reduction in death risk compared with people with a 'normal' BMI of between 18.5 and 25. Those with BMIs over 35, however, have a higher risk.[16]

It isn't clear why being overweight might protect against an early death. Perhaps carrying a few extra pounds in reserve helps the body fight off illness or infection. Perhaps overweight people are more likely to receive medical attention. Or perhaps some people in the normal category have become thin recently due to a serious illness.

Whatever the reason, the finding is not a green light to eat all the pies. Overweight people might be more likely to develop diseases that affect quality of life, for instance. Even so, it seems that a little bit of flab may not be the crime against health it has always been made out to be.

THE TRUTH ABOUT CALORIE COUNTING

A useful trick when trying to lose weight is to keep an eye on how much energy you are consuming, or even track it accurately. The standard measurement of energy in food is the calorie (technically the kilocalorie, kcal, though usually abbreviated to calorie).

Generally speaking men burn through about 2,500 kcals a day and women 2,000 – though this obviously varies from person to person and also depends on how much exercise you take. So a calorie-controlled diet needs to aim not to bust that limit.

Counting calories is a tedious business but if you want to lose weight and keep it off, it can be an indispensable discipline. One of the iron rules of nutrition science is 'CICO' – calories in versus calories out. If what goes in is greater than what goes out, you will gain weight. If out is greater than in, you'll lose it.

Sometimes a calorie is not actually a calorie, so *caveat emptor*. If you are one of those people who actually read food labels, this might make the difference between losing weight or gaining it. Prepare yourself for a shock: food labels are often wrong. Using the information on food labels to estimate calorie intake could be fraught with danger. Calorie estimates on food labels are based on flawed and outdated science, and provide misleading information on how much energy your body will actually get from a food. Some food labels may overestimate or underestimate this figure by as much as 25 per cent, enough to spike the guns of any diet, and over time even lead to obesity.

Calorie counts on food labels are based on a system developed in the late nineteenth century by American chemist Wilbur Olin Atwater. He calculated the energy content of various foods by burning small samples in controlled conditions and measuring how much heat they released. To estimate the proportion of this raw energy that was used by the body, Atwater calculated the amount of energy lost as undigested food in faeces, and as chemical energy in the form of urea, ammonia and organic acids found in urine, and then he subtracted these figures from the total. Using this method, Atwater estimated that fat provides 9 kcals per gram, carbohydrates and protein 4 kcals per gram, and fibre 2 kcals per gram. With a few modifications, these measurements of what is known as 'metabolisable energy' have been the currency of the calorie ever since.

Hence if you look at food packaging, the calorie count will be roughly nine times the mass of fat, plus four times the combined mass of carbohydrates and protein, plus two times the mass of fibre.

However, these values are approximate. Our bodies don't incinerate food, they digest it. And digestion – from chewing food to moving it through the gut and chemically breaking it down along the way – consumes energy, the amount varying from food to food. This can lower the number of calories your body extracts from a meal by anywhere between 5 and 25 per cent. Yet these energy costs are not reflected on any food label.

Dietary fibre is one example. As well as being more resistant to mechanical and chemical digestion than other forms of carbohydrate, it provides energy for gut microbes, and they take their cut before we get our share. These factors reduce the energy derived from dietary fibre by 25 per cent – down from the current estimate of 2 kcals per gram to 1.5 kcals per gram. Similarly, the true number of calories from protein is 3.2 kcals per gram, not 4. That's because it takes energy to break it down into its constituent amino acids.

Put into the context of real life, these relatively small corrections may make a big difference. Consider a chocolate brownie and a muesli bar. The brownie contains an advertised 250 kcals, while the muesli bar has more than 300 kcals. That may seem a surprise. But the label will overestimate the calories actually derived from the fibrous and protein-packed muesli bar, perhaps by enough to make it lower in calories than the brownie. When just 20 kcals per day more than you need can add up to roughly a kilogram of fat over a year, that's a big deal.

Errors in the Atwater factors for protein and fibre are not the only problem. The brownie will also be much softer in texture than the muesli bar, a factor that is known to lower the energy cost of digestion. When two groups of rats are fed diets that differ only in softness of the food pellets, the softies gain more weight, suggesting that texture matters – probably because of the effort of chewing the food. People whose diets contain more hard-to-chew foods tend to be thinner;[17] and experiments on reptiles fed raw or cooked carrots prove that raw ones require twice as much chewing.

What's more, the brownie is made from refined sugar and

flour, making it easier for our bodies to extract the available calories than from the complex carbohydrates of the oatmeal in the cereal bar. And while the Atwater system assumes that the proportion of food that passes through the gut undigested is more or less constant, at around 10 per cent, this is not the case. The body may excrete 30 per cent or more of coarse-ground wheat flour, while today's finely milled flours may be almost completely absorbed. As a result, foods made from these fine flours are likely to deliver practically all of their carbohydrate energy into the bloodstream.

Cooking, too, can soften food and hence increase how many calories the body can extract, another factor the Atwater system ignores.

Much of the energy from starch in plants, for example, is stored as amylopectin, which is semi-crystalline, does not dissolve in water and cannot be easily digested. Heat plants in water, though, and the crystalline forms begin to melt. The starch granules absorb water, swell and eventually burst. The amylopectin is shattered into short starch molecules called amylose, which is easily digested.

Cooking also makes meat more digestible. Proteins are like origami – intricately folded three-dimensional structures that stomach acids and enzymes can't easily access. Heat relaxes the proteins, exposing them to enzymes that chop up the amino acids so they can be absorbed into the bloodstream.

To explore to what extent cooking increases the caloric availability of food, you need pythons. That may sound like a strange choice, but pythons are useful models for studying digestion because they remain motionless for days after eating, making it easier to measure their energy budget.

One classic experiment tested the impact of cooking and

grinding food. The snakes were fed one of four meals: intact raw steak, ground raw steak, intact cooked steak or ground cooked steak. Grinding alone or cooking alone reduced the cost of digestion by around 12.5 per cent. Grinding and cooking together roughly doubled the decrease, to 23.4 per cent. That's a significant decrease in the cost of digestion.

A handful of human studies have also been done. Fed cooked or raw egg white, for example, people absorb 90 per cent of the calories in the cooked white but only 50 per cent from raw.[18]

Yet despite these large variations in how much energy the body has at its disposal either to use or store, none of this is reflected in the food labelling system, which leaves consumers in the dark about their dietary choices. So if food labels are giving consumers potentially misleading information, what should be done about it?

This is a question nutritionists have asked themselves, and the answer they came up with is 'nothing'. Back in 2002, the UN Food and Agriculture Organization (FAO) assembled an international group of scientists to investigate the feasibility of changing food labelling to reflect real calorie counts. The group decided to stick to the established method because, their report concluded, 'the problems and burdens ensuing from such a change would appear to outweigh by far the benefits'.

Nutritionists do acknowledge that the current system isn't perfect, but most argue that the Atwater system makes it easy to calculate a reasonably accurate calorie count. They also say that overhauling such an entrenched system would require a huge amount of research in animal models and human volunteers, plus a more complicated labelling system than consumers are used to, for little real public health benefit.

In any case, most of the complications produce calorie counts that are too high rather than too low. So if you are counting calories, you can probably cut yourself a little bit of slack now and then.

THE TRUTH ABOUT METABOLISM

We all know people who eat like horses yet are as thin as whippets, and others who seem to pile on the pounds simply by looking at food. While there is no doubt that the causal factor of weight gain is calories in exceeding calories out, it is clear that we are not all equally prone to weight gain.

The standard explanation for this is differences in metabolic rate, which holds that some people have faster metabolisms than others and thus find it easier to stay lean. There's some truth to this, but it is not quite as simple as it sounds.

The technical definition of metabolism is the totality of the chemical reactions occurring in the body at any moment. But when it comes to weight management, only a subset of these matter – the ones concerned with storing or using energy.

The seventeenth-century physician Santorio Sanctorius pioneered the study of metabolism. For three decades he recorded his own weight and that of the food and drink he consumed, and his urine and faeces. He found that for every 3.6 kilograms he consumed, he excreted just 1.4 kilograms. The rest, he concluded, was lost through his skin.

We now know that this differential is lost as heat when food is converted into energy. Any excess is squirrelled away into one of two energy stores: glycogen in the liver and muscle, and, when they are full, fat.

The central regulator of energy use is the thyroid gland in the neck, which releases hormones that speed up the rate at which cells generate energy – aka metabolic rate. This can affect your body shape. People with an overactive thyroid stay thin regardless of how much they eat. An underactive thyroid, on the other hand, causes loss of appetite but also weight gain. However, both conditions are rare, affecting only around one in a thousand people.

For everyone else the influence of metabolic rate is more subtle. You can measure your resting metabolic rate by spending twenty-four hours doing nothing much in a metabolic chamber, which measures how much heat you generate, oxygen you consume, carbon dioxide you breathe out and nitrogen you excrete. These measures can be used to calculate overall energy expenditure.

These experiments bust the myth that skinny people have faster metabolic rates. In fact, as a rule, the fatter you are the more calories you expend at rest. That is largely because you have more cells to service with energy. Larger people also have larger visceral organs, which are very energy hungry. A kilogram of heart or kidney tissue burns 440 calories a day, compared to just thirteen calories for a kilogram of muscle. And contrary to popular myth, muscle tissue at rest burns only slightly more energy than fat. Replacing one kilogram of fat with one kilogram of muscle increases energy demand by just nine calories a day.

In other words, innate differences in baseline metabolic energy demand are small – much smaller than calories burned during physical activity.

Nonetheless, we shouldn't dismiss the notion that some people are naturally more inclined to gain weight than others. In 'overfeeding' experiments, people given an excess of a thousand calories a day over and above their personal baseline demand invariably gained weight, but some gained more than others.[19] The difference can be as much as threefold.

Such studies demonstrate that there are individual biological differences in propensity to gain weight. These turn out to be overall fitness, muscle mass, testosterone levels, responsiveness to the satiety hormone leptin, and the tendency to preferentially burn fat rather than sugar as a fuel. In scientific parlance there are obesity-resistant people and obesity-prone people. You probably know from experience where you fall on the spectrum. Unfortunately there's not much you can do to change your obesity proneness, apart from getting fit.

THE TRUTH ABOUT HUNGRY GENES

Life really is not fair. Not only are some people born rich, beautiful or brainy (even all three), some are also born to be thin.

The cause of weight gain is, unequivocally, overconsumption of calories. But that does not necessarily mean it is driven solely by gluttony and laziness. We have seen that popular wisdom holds that some people have a fast or slow metabolism; another slice of folk wisdom is that some people have bad genes. And this is not entirely wrong: some people are genetically predisposed to gain weight.

In fact, genes may be responsible for as much as two-thirds of our variation in weight, but not in the way many people assume. The idea that some people have a genetically programmed slow metabolic rate is not without merit, but these genes are the minority. Instead, most 'fat' genes make people gain weight in a more insidious way: by subtly affecting how appealing food seems, and how quickly we feel full.

This is what is known as 'satiety responsiveness', which roughly equates to how likely you are to notice when you have eaten enough. It is strongly heritable, and can have a big impact on your susceptibility to putting on extra weight. If you are one of those people who can always find room for a dessert (or two), even after a heavy meal, chances are you have a low satiety response. Even small amounts of excess calories over years can lead to significant weight gain.

If genes are destiny, then are some people doomed to overeating? Not necessarily. Even if you have a naturally low satiety response experts say you can train yourself to increase it. Pay close attention to the ways you respond to food. Do you invariably clear your plate however much was on it? Do you regularly go back for more? Do you have 'trigger foods' that you cannot stop eating? If so, try paying closer attention to how full you're feeling, pausing halfway through a meal, for instance. Try to retrain yourself to eat smaller portions. Cook less and put the leftovers straight into the fridge. Serve up on a smaller plate and don't feel obliged to finish everything on it. At first you may crave more, but little by little your expectations, and appetite, will adjust.

You might also try keeping 'floodgate foods' out of the house, or build up resistance over time: if chocolate is a weakness, buy a small bar and carry it around but don't eat it. The idea is that over time you will become less tempted to indulge.

If you're still unconvinced by the benefits of denying yourself from time to time, consider the fact that poor diet is a leading cause of ill health and mortality, and that it is incredibly easy to slip into bad habits. Nutritionists call it 'mindless eating' – scoffing whatever you feel like, whenever you feel like it. Given our evolved physiology and the obesogenic environment we live in, this is an inevitable path to fatness and unfitness.

I regularly do a 16:8 fast and occasionally go on a low-calorie weight-loss diet. I find that the discipline of consciously denying myself food when I'm hungry, or cutting out junk for a couple of weeks, acts like a reset button on my mindless eating. Try quitting crisps for a week or two, and then eat a bag. You won't believe how fatty and salty they taste. In a good way, of course, but you'll notice how you were wolfing them down without realising just how fatty and salty they are. This is a reminder of how easy it is to become habituated to delicious flavours and allow our diets to spiral out of control.

That is just my experience, but the science suggests that periodically going on a diet can have all kinds of benefits. So go on, deny yourself. Life will be all the tastier.

THE TRUTH ABOUT VITAMINS AND SUPPLEMENTS

SOME SEE THEM as an insurance policy against a less than ideal diet. Others take them because they want even more of a good thing. Whatever the reasons, popping pills containing vitamins, minerals, oils, metal ions and more has become a routine part of a healthy lifestyle.

On a superficial level, this looks like a good idea. For all their failings, nutritional studies consistently show that people who eat plenty of fresh fruit and vegetables, whole grains, lean meat and fish, and unsaturated fats have lower rates of lifestyle-related diseases such as diabetes, cancer and heart attacks. Molecular components of that diet, such as antioxidants, vitamins, essential fatty acids and minerals, have been shown to be at least partly responsible for those health benefits. So why not extract these super-nutrients, distil them, put them in capsules and swallow them down? *Voilà*: a healthy diet without the hassle and expense of preparing elaborate meals. At the very least they can't do you any harm, right?

Wrong. In recent years serious doubts have been raised over whether nutritional supplements really do deliver goodness in a pill. Take omega-3 fish oil capsules. Once the darling of the supplement world, recent studies indicate that they do little good. Other studies have reached similarly deflating conclusions for the effects of an alphabet soup of vitamins. In some cases, they turn out to be positively harmful.

To understand the supplement craze, a bit of history can help. The seeds were sown at the end of the nineteenth century when scientists began to discover the nutritional roots of some common but hitherto mysterious diseases. For example, beriberi – a multi-symptom disease common at the time in eastern Asia – could be prevented by feeding people brown rather than white rice. In 1911 Polish chemist Casimir Funk isolated an organic compound from rice husks that turned out to be responsible. It was an amine, so he called it 'vitamine', a contraction of 'vital amine'. He proposed that similar compounds would be found that could prevent pellagra, scurvy and rickets, and was proved right. We now know the anti-beriberi compound as vitamin B1 and the others as vitamins B3, C and D.

The following two decades saw many more vitamins and other micronutrients discovered and a growing understanding of how a lack of them caused common illnesses. This led to dietary strategies that cured all manner of previously incurable diseases. Science now recognises around a dozen vitamins, plus numerous compounds including fatty acids, amino acids and minerals, that must be present in the diet because the human metabolism cannot make them. These are termed 'essential'.

The success at identifying, preventing and curing nutritional deficiencies naturally led to the idea of dietary supplements and fortification. In the US, iodine was added to table salt in 1924 to prevent goitre, vitamin D to milk in 1933 to prevent rickets, and several vitamins and minerals were added to flour in 1941.

Public awareness of vitamins grew, and with it a common-sense desire to take advantage. Vitamins turned out to be very cheap and easy to formulate into tablets. Single vitamin supplements went on sale in the US in the 1930s and multivitamins a decade later. Other micronutrients followed and today there is a mind-boggling offering of supplements, both singly and in elaborate combinations, to choose from.

An estimated half of the US population and nearly a third of people in the UK take some form of supplement each day. All those pills add up to a very profitable enterprise: the industry is worth roughly $50 billion.

Supplements often have the aura of validated medicines, but they are not. To be approved for sale they face much lower regulatory hurdles. Pharmaceuticals have to jump through the hoops of clinical trials, proving exhaustively and very expensively that they do what they say they do, over and above the placebo effect. In contrast, supplements are generally regulated in the same way as foodstuffs. They need to be proven safe for human consumption and correctly labelled, but they don't have to prove scientifically that they have health benefits.

Dietary supplements cannot make specific health claims, but as a quid pro quo their marketing people have cannily found ways to get the message across anyway. Bear that in mind when choosing a supplement. Also consider that 'safe for human consumption' doesn't necessarily mean safe for long-term human consumption. Any studies that are done tend to be small and brief, which means they don't pick up subtle or chronic effects.

As for the logic of extracting and distilling individual micronutrients into pill form, this turns out to be seriously flawed. People don't eat micronutrients in isolation but with a mixture of whole foods. How much you eat, when you eat it and in combination with what can have a huge impact. In one recent experiment, for example, people were fed fresh fruit and vegetables either on their own or in combination with fats and spices. Far more of the nutrients entered the bloodstream when the extras were included.

This is known to happen in the real world. Olive oil, for example, increases the amount of a healthy compound called lycopene that we absorb from tomatoes, because carotenoids must be dissolved in fat to be transported into the blood. A similar principle applies to the fat-soluble vitamins A, D, E and K.

This could be why supplementation sometimes backfires. For instance, dietary calcium seems to reduce people's risk of kidney stones, but calcium supplements have been linked to an increased risk.

Another worry is that healthy dietary compounds extracted, purified and delivered at high, concentrated doses may be harmful. The plant pigment beta carotene, for instance, is an important component of a healthy diet (it is converted to vitamin A in the body and is one of the most important dietary sources of that vitamin) but supplements increase the risk of lung cancer in smokers. A surfeit of vitamin E has been linked to a greater risk of stroke, and possibly prostate cancer (interestingly, both vitamins A and E are known to be dietary antioxidants; we'll get onto antioxidants on page 149).

Because of the way in which supplements are regulated, we have essentially invented a gigantic system of self-diagnosis and self-medication. Are you really qualified to decide that you are deficient in some nutrient – one of hundreds found in a balanced diet – and prescribe yourself a pill to resolve it? What people take is based more on magical thinking, advertising and word of mouth than science.

As a rule, whole foods are the optimum source of the nutrients we need. If you eat a mixed and healthy diet it is highly unlikely that you actually need to supplement it with vitamins, minerals or other micronutrients. Recall that the supplement industry has its roots in curing genuine dietary deficiencies.

Some nutrition researchers argue that for people who genuinely find it hard to follow a balanced, omnivorous diet, perhaps because of allergies, lifestyle choices such as being vegetarian or vegan, or simply disliking certain foods, taking supplements may be helpful. For those with a poor diet, topping up with supplements can bring them closer to the recommended daily dose. But supplements are not the ideal substitute. Mainstream nutrition science is moving towards the position that, unless you have a specific deficiency, routine supplementation is pointless at best and may even be risky. A qualified professional is best placed to diagnose dietary deficiencies.

For people who do decide to carry on taking supplements, there are some simple things you can do: always take them with food, including a little fat, and follow the instructions on the packaging to avoid overdosing. And consult your doctor about any supplements you are taking, as some vitamins have been shown to interfere with drugs, such as the blood-thinning medication warfarin.

Most of all, don't assume that taking nutritional supplements can compensate for an otherwise poor diet. There are thousands of active ingredients in foods, some of which we are only starting to become aware of. The best route to a healthy diet is to eat one.

THE TRUTH ABOUT SUPPLEMENTS, FROM VITAMIN A TO ZINC

When it comes to supplements, it's useful to know that vitamins can be separated into water-soluble and fat-soluble varieties. The B vitamins and vitamin C are the water-soluble kind. They are absorbed from food until they reach a saturation threshold, beyond which any excess is urinated out. You can't store them, so if you stop consuming them your levels quickly fall.

The fat-soluble vitamins A, D, E and K, on the other hand, can be stored in the body, particularly in the liver. That means you don't need to consume them every day.

With all of this in mind, which supplements – if any – are worth the money?

Vitamin A

Vitamin A is an umbrella term for several compounds including retinol and retinoic acid. It is vital for vision, immune function and healthy skin. Daily requirements are 0.6 milligrams for women and 0.7 milligrams for men. The NHS advises that you should easily get enough from food and don't need to supplement your diet.

Dairy products, fish oils and liver are the richest sources. Fruit and vegetables contain precursor forms, notably the orange pigment beta carotene, which get converted into vitamin A in the body. Just half a carrot meets your daily requirement.

Some supplements provide high doses, and we store the surplus away in fat, where it can build up to harmful levels. An intake of more than 1.5 milligrams a day on average may interfere with the beneficial effects of vitamin D and weaken bones. Even eating too much liver can expose you to vitamin A poisoning, called hypervitaminosis A. Polar bear liver is famously toxic because it stores so much vitamin A. Walrus and moose livers are similarly dangerous. In the unlikely event that you are offered one, politely decline.

But don't worry about overdosing on carrots. The conversion of beta carotene to vitamin A is tightly regulated and your body won't make more than it needs. However, beta carotene supplements are a different story. In people who smoke or have been exposed to asbestos, they increase the risk of lung cancer.

Speaking of cancer, in the 1980s epidemiologists speculated that high doses of vitamin A might protect against it. But several well-designed trials, and a recent meta-analysis, have failed to show any benefit.[1] In fact, they suggest the precise opposite. Vitamin A supplements may increase the risk of cancer.

VERDICT: Potentially harmful. Avoid vitamin A supplements unless you have a diagnosed deficiency. If you smoke, avoid beta carotene supplements.

B vitamins

If you are a fan of vitamin B, are you ready to B heartbroken?

There are eight B vitamins – 1, 2, 3, 5, 6, 7, 9 and 12 – all of which are essential for metabolism. But they are chemically distinct and have quite different functions. They are grouped together for the simple reason that they are often found in the same foods.

Some members of the family are better known by their chemical names: thiamine (B1), riboflavin (B2), niacin (B3), biotin (B7) and folate/folic acid (B9). The 'missing' numbers (4, 8, 10, 11 and 13–20) were assigned to compounds once thought to be essential in the diet but since discovered not to be.

The B vitamins are water-soluble and surpluses are excreted in urine, so we need to eat them every day to avoid deficiencies. The two best known are beriberi (B1 deficiency) and pellagra (B3 deficiency), though these only appear after prolonged and severe malnourishment. Well-nourished people can easily get all the B vitamins needed from their daily diet. Vegans are the exception, as B12 is found naturally in large quantities only in meat, eggs and dairy products; B12 deficiency causes skin lesions and people who choose not to eat animal products are advised to take supplements or eat B12-fortified foods such as breakfast cereals.

THIS BOOK COULD SAVE YOUR LIFE

People over the age of fifty should also consider B12 supplements. As we age, we produce less stomach acid, which can impede our ability to absorb B12 from our diet. Older adults are often deficient in this nutrient.

Many people who do get enough B12 from animal products choose to take other B vitamins as well. Individual compounds are available, and so are supplements containing all eight, called vitamin B complex. Supplements of B vitamins are claimed to help prevent heart disease, stroke and cognitive decline. But randomised trials have not supported this claim.[2] Excess quantities are simply peed down the toilet, so you are wasting your money.

But even though you excrete the surplus, excessive doses of some B vitamins are not harmless. Taking too much nicotinic acid (a form of niacin, B3, which you need about fifteen milligrams of a day) can damage your liver. Prolonged consumption of more than 200 milligrams a day of vitamin B6 can lead to numbness in the arms and legs known as peripheral neuropathy. It usually goes away after you stop taking the pills but severe cases, caused by several months of excessive B6, can be permanent.

VERDICT: For vegans, vegetarians who eat very little dairy and older adults, B12 supplements are a good idea. For omnivores, they and all other B vitamins supplements are unnecessary.

Vitamin C

C is the most popular single vitamin supplement in the US. As well as its antioxidant properties (see Vitamin E), it supposedly wards off colds and protects you from cancer. But unless you're doing subarctic military training, a daily vitamin C tablet will not prevent colds.[3] The evidence for cancer prevention is similarly lacking.

Lack of vitamin C is the cause of perhaps the most famous and feared deficiency of all, scurvy. Vitamin C helps us absorb iron and make collagen, a protein that provides a scaffold for skin and helps wounds heal. Without enough, you succumb to bleeding gums and weeping wounds. Adults require at least forty milligrams each day, which is easily achievable by eating citrus fruits, peppers, broccoli and strawberries. One orange will supply your daily needs.

So how did taking vitamin C to prevent colds become so popular? We can thank the Nobel Prize-winning chemist Linus Pauling, who became obsessed with it as a panacea and advocated taking colossal doses. The evidence is weak, however: a recent review found that taking 200 milligrams of vitamin C every day may reduce the severity and duration of a cold, but it won't prevent it in the first place – at least not unless you're a soldier training in extreme cold, a marathon runner or a competitive skier.

Taking lots of vitamin C is unlikely to harm you, but mega-doses may cause diarrhoea, heartburn and other problems. The upper recommended limit for daily intake is 1,000 milligrams.

Pauling also thought vitamin C could help cure cancer. There is some anecdotal evidence that people with cancer given high doses intravenously survived longer than expected, but it is impossible to say whether that was because of the vitamin. Higher-quality research into its preventive effects is more damning: large-scale trials have found no evidence of reduced cancer risk.

VERDICT: It won't prevent colds but if you've caught a cold, vitamin C may help you recover faster. Otherwise, it's worthless.

Calcium

Half of US adults take some form of calcium supplement, largely in the belief that it will strengthen their bones and teeth. It is certainly true that calcium is important for strong bones and teeth, and that severe dietary deficiency may lead to bone fractures and osteoporosis. For good bone health you need at least 700 milligrams of calcium each day, rising to 1,200 milligrams after the age of fifty.

But supplements are a very bad way to get calcium. A number of large studies have looked at whether calcium supplements reduce the risk of fractures but found nothing to suggest that they do. At the same time, evidence has emerged of serious health risks.

Too much calcium causes calcification of the arteries, and you should aim to eat no more than 2.5 grams a day. You'd have to go some to consume that much – it's about what is in 350 grams of cheese.

But supplements can deliver very high doses, and bring their own dangers. They increase the incidence of kidney stones by 17 per cent and can cause gastrointestinal problems so severe that they require hospitalisation – mostly constipation but also cramping, bloating, pain, vomiting and diarrhoea. A bigger concern is risks to the heart. A five-year trial of post-menopausal women found that those taking calcium supplements were more likely to suffer a heart attack or stroke than those who didn't.[4] Subsequent research suggests that the risk of a heart attack rises by up to 40 per cent among people taking calcium supplements.[5] Each of these alone is enough to outweigh any of their benefits. Taken together they are a big red flag: take calcium supplements at your peril.

Fortunately you can easily get what you need from your diet.

Dairy products are a great source, as is oily fish. People who are vegan or lactose intolerant may worry about getting enough, but can get plenty from green vegetables, okra, figs, chia seeds and almonds. Many foods are fortified with calcium, and current knowledge suggests that eating them is equivalent to consuming calcium in food rather than in supplements.

VERDICT: Don't.

Chromium

Whether or not chromium is a dietary essential is hotly debated. Its biological function in humans has not been firmly established, and may be non-existent. Health authorities in the US, Australia and New Zealand regard it as essential, but the European Food Safety Authority does not.

The claim that we need it is based on unexpected weight loss and other health problems seen in people on long-term intravenous diets containing no chromium. In places where it is considered essential, the 'adequate intake' is twenty-five to thirty-five micrograms a day for adults.

Millions in the US take chromium supplements, probably lured by claims that it can burn fat, help control blood sugar and raise good cholesterol, and hence lower heart attack risk. But these are far from established. A recent review of studies found 'no current, reliable evidence' that chromium can help overweight adults slim down.[6] It is a similar story for glucose control. The cholesterol claim also lacks solid evidence.

According to the NHS, if you really want to take it, 10 milligrams a day or less should be fine. But why bother when you can get all you may (or may not) need by eating fruit and vegetables? Broccoli is the richest dietary source. One small portion will fulfil your daily quota.

VERDICT: Unless you really can't stomach fruit and vegetables, don't waste your money

Coenzyme Q10

The many claims made by proponents of coenzyme Q10 (CoQ10) – that it can help treat heart failure, cancer and diabetes, improve sperm quality and even help you live longer – stem from the fact that the compound plays a crucial role in almost every cell in the body.

In small studies, CoQ10 has been shown to boost immune function in people with cancer and HIV, prevent heart damage caused by some chemotherapy drugs, and improve heart function in people recovering from heart failure. But other claims fail to stand up to scrutiny. An analysis by the European Food Safety Authority dismissed most as lacking in evidence, including claims that CoQ10 can increase energy.[7]

So how much do you actually need? There are no recommended daily intakes, but one study found that adults consume about three to five milligrams per day. Most of us get all the CoQ10 we need from our own bodies and from what we eat: liver and oily fish are good sources.

Deficiency is rare and usually associated with neuro-degenerative disorders, conditions such as kidney disease or genetic mutations that prevent the body from making the coenzyme.

VERDICT: For now, more mouth than trousers.

Vitamin D

Vitamin D is unique among vitamins in that we can make our own, which rather undermines its claim to be a vitamin. But that doesn't mean that supplements are a total waste of time. In fact, unlike most other vitamins, there is sometimes a decent case to be made for taking supplements.

We make vitamin D through the action of sunlight on our skin. In the summer, light-skinned people can make more than they need each day by spending five to ten minutes in the sun with some skin exposed. Darker-skinned people need about thirty to forty minutes.

But in wintertime you might struggle to make enough. If you live north of San Francisco or southern Europe, or south of Melbourne, Australia, the winter sun isn't high enough in the sky even at midday. The rays that matter, UVB, don't reach the surface of the Earth.

Our bodies can store vitamin D and it is possible to stash enough away during the summer to get you through the darker months. However, about 40 per cent of adults in the UK and US are deficient in vitamin D during the winter, which may be impacting their health. People with very low levels of vitamin D are at 57 per cent greater risk of death from all causes than people with high levels.[8]

Vitamin D deficiency causes the bone disorder rickets, but has also been linked to an increased risk of autoimmune diseases and greater susceptibility to viral infections.

Do we need supplements? Most guidelines agree that young children, pregnant women and elderly people should take them – and possibly adults who don't get out in the sun much. The UK's Scientific Advisory Committee on Nutrition recommends that everyone should consider taking vitamin D supplements during winter. You can also get vitamin D from food; tuna, salmon and eggs are the best sources.

Recommended daily intake is about fifteen micrograms, but experts say up to a hundred micrograms per day is probably safe. Too much vitamin D can cause calcium to build up in the blood, which can result in vomiting and kidney problems.

Vitamin D tablets are a reasonably effective substitute but for unknown reasons do not seem to confer the full range of health benefits obtained from sunlight or food.

VERDICT: Pregnant women and the elderly are advised to take supplements. If you don't fall into these categories you might still benefit, but only if you live at a higher latitude than the 38th parallel (north or south), it is winter, you didn't get much sun last summer and you eat a diet that doesn't contain much oily fish.

Vitamin E

Vitamin E is the quintessential antioxidant, which means it can neutralise toxic free radicals, at least in a test tube. Radicals are an unavoidable by-product of cellular metabolism and cause oxidative damage to our cells, organelles and biomolecules.

For this reason it was once assumed that taking lots of vitamin E (ditto vitamins A and C, and selenium) in supplement form should help prevent cancer and other diseases associated with free-radical damage. But the evidence doesn't stack up. Indeed, in high doses vitamin E supplements appears to increase the risk of prostate cancer (for more on antioxidants see page 149).

We do need vitamin E to help prevent blood clots and maintain immune defences. Deficiency can lead to muscle weakness and vision problems, and over-prolonged periods can damage your nerves, liver and kidneys.

Vitamin E deficiency in otherwise healthy people is rare – even getting slightly less than you need doesn't appear to be harmful, largely because we can store it in fat. The daily intake recommended for adults ranges from seven to fifteen milligrams a day. You should easily get that from your diet. Rich sources are vegetable oils, nuts, seeds and green leafy vegetables.

With regard to other conditions supposedly prevented by vitamin E supplements – including heart disease and dementia – there is no evidence of benefit. Also, despite the popular belief that vitamin E applied as a cream or gel can make scar tissue fade, research doesn't bear this out.

VERDICT: Potentially harmful. Don't take vitamin E supplements unless you have been diagnosed with a deficiency, probably caused by another condition.

Fish oils

The oil business is booming; fish oil is the second most popular dietary supplement in the UK after multivitamins. These shiny golden capsules contain vitamins A and D but are most prized for delivering that precious class of fatty acid, omega-3s (see page 18).

Omega-3s are certainly vital. They are a key ingredient of cell membranes and have numerous health-giving properties including protection against cardiovascular disease. Omega-3s cannot be synthesised in the body so must be obtained from our diet. They are indelibly associated with oily fish, but the most important one – alpha-linolenic acid (ALA) – is not actually found in fish oil. The best sources are chia seeds, kiwi fruit, walnuts, flax seeds (linseed), rape (canola), soybean oil, seaweed and leafy greens.

However, fish oils do contain the two other important omega-3s: eicosapentaenoic acid (EPA) and docosahexaenoic acid (DHA). The oil in the capsules usually comes from cold-water oily fish such as anchovies, herring, mackerel, salmon and sardines.

For all three omega-3 fatty acids, average intake among adults in the US and UK falls far short of the recommended amount, largely due to the fact that many people eat little or no oily fish. Taking fish oil supplements will address deficiencies of EPA and DHA but – disappointingly, given their golden reputation – doesn't seem to do anything for cardiovascular health.

A recent meta-analysis pooled the data from seventy-nine studies to assess the long-term impact of omega-3 fatty acids or fish oil supplements on cardiovascular health. It concluded that they have little or no effect.[9] That may be because they don't contain ALA, or for some other unknown reason.

VERDICT: It's probably best to get your omega-3s from food, but if you don't like oily fish, supplements can supply two of the vital three. But make sure you also eat sources of the other one, alpha-linolenic acid (ALA).

Folic acid

The possession of folic acid supplements – also called folate or vitamin B9 – is often taken as a telltale sign that you're trying for a baby or are already pregnant. And with good reason: they definitely prevent spina bifida and other spinal deformities that occur in the womb. But other people also take folate for their cardiovascular health. Is it worth it?

Folic acid is important for the formation of red blood cells, and may also be involved in the production of DNA and RNA. Adults need 200 to 400 micrograms daily.

Before and during pregnancy the recommended intake rises to 400 to 600 micrograms per day, or more for women with a family history of foetal spinal abnormalities. These are generally caused by a defect in the neural tube, a structure that forms in early pregnancy. Taking folic acid before conception and in early pregnancy reduces the risk of neural tube defects by as much as 70 per cent.

More than seventy countries – including the US, Canada and Australia, although not the UK – require that flour and cereals are fortified with folic acid. Since fortification became widespread in the US in 1998, the number of babies born with a neural tube defect has dropped. But fortification does not eliminate the problem – neural tube defects still affect roughly 3,000 US pregnancies a year – so pregnant women are still advised to take supplements.

There is less evidence to back general consumption of folic acid supplements. A balanced diet should supply all you need; leafy greens, salmon and beans are good sources. An analysis of several studies found that daily supplements did not reduce rates of heart disease or stroke.[10]

High doses of folic acid – more than one milligram – can hide the symptoms of vitamin B12 deficiency but don't compensate for it. B12 deficiency can eventually lead to nervous system damage.

VERDICT: Supplements are advised during pregnancy. Otherwise, you should get enough from food.

Glucosamine

The sugar derivative glucosamine is a building block of cartilage and a component of the fluid around joints. No surprise, then, that it is often touted as a remedy for arthritis. In animal studies, glucosamine can decrease pain and inflammation and slow the erosion of cartilage. Study methods vary hugely, though, making it difficult to say anything conclusive.

Small trials in humans suggest a daily pill can reduce arthritis pain and inflammation, but meta-analyses find little difference between glucosamine and placebo.[11] That said, they also find no difference in harm between the two. So glucosamine may not help much, but is unlikely to make things worse.

Still, the UK's National Institute for Health and Care Excellence (NICE) tells doctors not to recommend glucosamine for arthritis because of the lack of convincing evidence and the potential risk factors to people with shellfish allergies, hypertension or diabetes.

For these people it is a no-no. Glucosamine supplements are often derived from the shells of shellfish, some formulations contain a lot of salt which can add to hypertension, and there is some evidence that glucosamine can hinder the body's ability to process glucose.

VERDICT: For most people, taking glucosamine to tackle arthritis won't hurt – but there's not enough evidence to say it will necessarily help either.

5-HTP (5-hydroxytryptophan)

An amino acid produced in the body, 5-HTP regulates levels of the neurotransmitter serotonin in the brain. Proponents of 5-HTP supplements claim they can increase energy and even treat depression. There is some evidence for the latter claim, at least, but 5-HTP may not be worth the risk.

An analysis of 111 published trials of 5-HTP found just two that met rigorous standards.[12] These suggested it may be better than placebo for depression, but the researchers would not recommend it: they also noted that the supplements may be linked to a very rare but potentially fatal syndrome known as eosinophilia-myalgia, a condition that causes incapacitating pain. 5-HTP may also throw other brain chemicals out of whack and lead to side effects such as anxiety.

VERDICT: There isn't enough evidence to prove that 5-HTP does any good, but there is enough to suggest it may be harmful.

Iodine

Iodine is an essential micronutrient that has one function and one function alone in the human body: as a component of the hormones thyroxine and triiodothyronine, which are made by the thyroid gland in the neck. Both are important regulators of metabolic rate.

Daily requirements are very small, just 0.14 milligrams. The only reason to take iodine is to address a deficiency, but it is extremely unlikely that you are iodine deficient. A balanced diet supplies plenty and iodine is routinely added to table salt to fill any gaps. Taking up to 0.5 mg a day won't do you any harm, except as a needless drain on your wallet.

VERDICT: You almost certainly don't need it.

Iron

The human body supposedly contains enough iron to make a three-inch nail. It is a major component of haemoglobin, the protein that carries oxygen in the blood, and myoglobin, which does the same in muscle. It is also found in various enzymes involved in metabolism, immune response and the production of neurotransmitters. So iron is vital to our health.

However, it should be easy to get the recommended daily adult intake of eight milligrams (fifteen for pre-menopausal women) from your diet, as many common foods contain oodles of iron. Meat and fish are rich sources, as are pulses, dark green leafy vegetables, and cereals and bread fortified with iron.

The only compelling reason to take iron supplements is if you have iron-deficiency anaemia, which needs to be diagnosed by a doctor – not least because it can be a symptom of something more serious, such as cancer or a stomach ulcer.

The classic symptom of iron-deficiency anaemia is tiredness and lack of energy. But having too much iron in your system, which is called iron overload, also causes these symptoms. Some people are at elevated risk of this: heavy drinkers, people with inflammatory conditions like rheumatism, carriers of certain genetic mutations – and those who swallow too many iron supplements.

Excess iron is highly reactive, forming free radicals that damage cells, leading to type 2 diabetes and certain cancers. An acute overdose of iron can be fatal; young children are especially vulnerable.

So if you think you need an iron supplement, first try upping the iron in your diet and cut down on tea, coffee, milk and whole grains, which can prevent your body from absorbing iron from your diet. If that doesn't work, visit your doctor.

VERDICT: Before taking iron supplements, talk to a doctor: too much iron is at least as dangerous as too little.

Vitamin K

Vitamin K supplement sales are booming thanks to the promise that it can strengthen your bones. That may be true, but the evidence is not strong.

Vitamin K comes in two forms: K1, found in leafy green vegetables; and K2, synthesised by bacteria in the gut. It is best known for its role in blood clotting (its name is derived from its original German title, *Koagulationsvitamin*). The very rare cases of deficiency in adults are associated with uncontrolled bleeding. It can interfere with blood-thinning medications, so doctors recommend against supplements for people taking these.

Requirements are minuscule, roughly one microgram per kilogram of body weight a day. At this point you will be unsurprised to hear that you can obtain plenty from food without making a special effort. Green leafy vegetables, vegetable oils and grains are good sources.

There are a handful of small studies of vitamin K supplements, and not all point the same way: some suggest it promotes bone health, while others find it makes no difference. The negative effects of excessive doses, if any, are not known.

VERDICT: Just eat some broccoli.

MSM

MSM (methylsulphonylmethane) is a sulphur-containing compound found in meat, vegetables and other foods. Advocates of MSM supplements claim it is a good source of 'biologically active sulphur', but that is a biologically meaningless term.

Sulphur is an important component of many proteins and hence important in our diets, but deficiency is unknown as most foods are stuffed with protein.

MSM supplements have been touted as useful against ailments from pain to cancer and brittle nails. However, very little published research has put MSM to the test. Most of what has been done focuses on osteoarthritis and has generally found no effect.[13] So while MSM is not thought to be harmful, there seems little point in taking it.

VERDICT: Sulphur smells of rotten eggs, and so do the health claims for MSM.

Multivitamins

They are marketed as 'dietary insurance', and each year multivitamins bring in $5.2 billion in the US. But the evidence suggests this is mostly money down the drain. If in doubt, reread the sections on vitamins A, B, C, D, E and K (and iron, which is frequently added to multivitamins), add them all together, and draw your own conclusion. (Spoiler alert: the US National Institutes of Health sees no reason to recommend them.)

Having said that, people with conditions that affect absorption of essential nutrients, such as Crohn's disease, may benefit.

A recent trial did point to a slightly decreased cancer risk for men taking multivitamins. But it found no impact on cardiovascular health, and the results of many other studies looking at possible benefits are a total muddle. Experts echo this point of view, adding the concern that, beyond being ineffective, supplements that contain high doses of vitamins A and E and iron may even cause harm.

VERDICT: Not beneficial except in rare diseases and, depending on the levels of vitamins A and E and iron, may be harmful.

Magnesium

Many enzymes require magnesium to function properly, including those involved in nerve and muscle function and in regulating blood glucose and blood pressure.

You need about 300 milligrams a day. A balanced diet should do the trick. Nuts and spinach are good sources. But teenagers and people over seventy often get too little, the former due to a diet high in junk food the latter an age-related decline in absorption efficiency. Diabetes and certain medications can also lead to deficiency.

However, it is rare for magnesium deficiency to cause illness, and beyond treating extremes the evidence on the benefits of supplements is sketchy. Up to 400 milligrams a day should be fine, but any more can lead to diarrhoea; some laxatives contain large amounts of magnesium. Mega-doses – more than five grams per day – can cause difficulty breathing and even a heart attack.

Magnesium is touted as a way to avoid diabetes, cardiovascular disease and migraine. But the evidence is not strong enough to support these claims.

VERDICT: Unless you have a diagnosed deficiency, don't waste your dough.

Selenium

Selenium is a trace element found in fish, meat, grains and dairy products. It is essential for the synthesis of selenoproteins, a small but vital class of proteins. An adult only needs about sixty micrograms per day, which is easy to get by eating meat, fish or nuts. In the West selenium deficiency is very rare.

Selenium is often touted as an antioxidant and hence protection against cancer, age-related cognitive decline and cardiovascular disease.[14] But antioxidant supplements don't work (see antioxidant myths, page 149).

Selenium is also said to help with asthma, reduce the risk of rheumatoid arthritis and slow the ageing process. Even if these claims are true, there is no compelling reason to take a supplement, even if you don't eat meat or fish. One Brazil nut contains more than your daily needs.

Selenium can be poisonous in large doses, causing a condition called selenosis, which makes your hair and nails fall out, and can be fatal. The NHS says don't take more than 350 micrograms a day. Most supplements contain 50 to 200 micrograms.

VERDICT: Buy a bag of Brazil nuts instead.

Potassium

Potassium works in tandem with sodium to regulate the volume of blood in our body, and helps us get rid of excess salt – and with it, excess fluids – through urine.

High blood pressure is a major cause of heart disease. And the popularity of processed foods, which tend to be high in salt and low in potassium, adds to the problem. In healthy people, the ratio of sodium to potassium should be roughly 1:1. So how do we get the 3.5 grams per day recommended by the WHO? For most of us, it's down to – you guessed it – eating fresh fruit and vegetables.

Still, supplements are popular, but in otherwise healthy people potassium deficiency is very rare. Supplements may pose a risk for people with high blood pressure or certain kinds of kidney disease that mean they cannot get rid of potassium, but otherwise are unlikely to cause harm.

VERDICT: For most people potassium supplements aren't worth it. Eat a banana instead.

Zinc

In a classic episode of *The Simpsons*, Bart and his schoolmates watch an educational film called 'A World Without Zinc'. For reasons unexplained, a man called Jimmy wants to live without zinc. His wish is granted, but he soon regrets it: he can't go on a date because his car won't start, and he can't call his girlfriend because his telephone won't work. Horrified at what he has done, he tries to shoot himself. But his gun won't fire, because the firing pin is made of zinc.

A real-life world without zinc would be deadly for another reason. Zinc is an essential micronutrient. It plays a vital role in cell division, is required for about a hundred enzymes to work, and helps with metabolism, wound healing and the production of DNA.

The recommended daily amount is 7 milligrams for women and 9.5 for men. Too little can result in a weakened immune system, hair loss and mental slowness. At this point you will be flabbergasted to learn that you can get that easily from food. Oysters are a rich source; so are meat, cheese and wholegrain wheat.

But you may also be surprised to learn that there is – sometimes – a good reason to take zinc supplements. There is decent evidence that if you have a cold, taking a daily supplement shortens its duration, as long as you start within twenty-four hours of falling ill.[15] But if you don't have a cold there is no good reason to take zinc; if you do, don't take more than twenty-five milligrams a day, to avoid the risk of anaemia and weakened bones.

VERDICT: Good for the common cold, useless for everything else.

THE TRUTH ABOUT ANTIOXIDANTS

Multivitamins. Cranberry capsules. Green tea extract. Vitamin E. Pomegranate concentrate. Beta carotene. Selenium. Grape seed extract. Royal jelly.

You name it, if it's touted as an antioxidant, we'll swallow it. An estimated one in two adults in the US take one or more daily. But it turns out we may have swallowed a myth.

Antioxidants acquired a reputation as miracle health supplements thanks to research started in the 1950s, when scientists discovered that many diseases – including heart attacks, strokes, cancer, diabetes, cataracts, arthritis and neurodegenerative disorders – were linked to damage caused by highly destructive chemicals called free radicals.

Free radicals are compounds with unpaired electrons that stabilise themselves by ripping electrons from other molecules including proteins, carbohydrates, lipids and DNA. In the process they often create more free radicals, sparking off a chain of destruction. Oxidative damage accompanies most, if not all, diseases and has even been proposed as a leading cause of ageing.

Free radicals are an unavoidable hazard of being alive. Around 1 per cent of the oxygen we consume turns into radicals, meaning that the human body generates 1.7 kilograms a year. Ozone, air pollution, tobacco smoke, sunlight, microbial infections and industrial chemicals also trigger free radical production.

In the 1980s, however, a potent weapon against free radical damage appeared. It had been known for a long time that people whose diets are rich in fruits and vegetables had a lower incidence of those diseases that are associated with free radical damage. Now there was an explanation. Fruits and vegetables are a rich source of antioxidants that can neutralise free radicals by donating electrons to them. Plants are full of antioxidants for good reason. They are especially vulnerable to oxidative stress since they produce pure oxygen during photosynthesis.

And so a lucrative hypothesis was born: dietary antioxidants are free radical sponges that can stave off the diseases of old age. Extract them, purify them and sell them as supplements and everybody can reap the rewards of a vegetable-rich diet without the actual vegetables.

The best-known antioxidant supplements are vitamin A, vitamin E (also known by its chemical name tocopherol), vitamin C, and two broad classes of plant chemicals called polyphenols (including flavonoids) and carotenoids (including beta carotene

THE TRUTH ABOUT VITAMINS AND SUPPLEMENTS

and lycopene). Most supplements touted as antioxidants contain at least one of these, often as a pure chemical or sometimes as a concentrated plant extract.

It was a highly plausible and reasonable hypothesis. But like many such hypotheses it did not survive contact with reality.

As food supplements, antioxidant pills did not have to pass clinical trials to be marketed; basic safety tests were all that were needed. But scientists decided to do clinical trials anyway, if only to prove that antioxidants worked as expected.

In the 1990s, the results of trials of some of the most popular supplements, including beta carotene, vitamin E and vitamin C, started to come in. They consistently and bafflingly found that while these substances are powerful as antioxidants in the test tube, taking them in pill form does not provide any benefit.

In fact, some studies suggest that they are actively harmful. A 2007 review of nearly seventy trials involving 230,000 people concluded that antioxidant supplements do not increase lifespan, and that some – beta carotene and vitamins A and E – actually decrease it.[16] When the results were published in the prestigious *Journal of the American Medical Association*, many people refused to believe them.

But no evidence since has given scientists reason to revise that conclusion. For example, in a prostate cancer prevention trial in 2011, men who took vitamin E for five and a half years had a 17 per cent greater risk of developing the disease than men who took a placebo.[17] It is now generally accepted that the supposed benefit of antioxidant pills is a medical myth.

Exactly why they don't work remains a bit of a mystery. Foods rich in antioxidants, such as fruit and vegetables, definitely protect against the diseases supposedly caused by radical damage. It may be that the antioxidants in these whole foods are more bioavailable or bioactive, and that extracted and purified anti-oxidants are stripped of their power.

Alternatively, the health benefits of vegetables may have little or nothing to do with antioxidants. Maybe people who eat them have a generally healthier lifestyle. Or maybe vegetables are beneficial because the antioxidants and other phytochemicals they contain are mildly poisonous and activate our own internal protective mechanisms – a process called hormesis (for more on this, see page 255).

But these hypotheses do not explain why some antioxidant supplements can be harmful. There are multiple potential reasons why this might be the case. One idea is that high levels of free radicals stimulate cells to ramp up their own antioxidants. These internal defences are far more effective than the antioxidants we swallow. So by taking supplements we may be deactivating a first-rate defence mechanism and replacing it with a second-rate one.

Another idea is that our immune systems harvest free radicals and weaponise them to kill bacteria and cancer cells. Antioxidants from supplements may therefore blunt our internal defences, and hence be counterproductive.

Yet another possibility is that antioxidants interfere with the beneficial effects of exercise, as there are hints that the body might use free radicals to prevent cellular damage after exercise.

Despite this solid and growing body of evidence that the supplements don't work, the antioxidant juggernaut rolls on. Global sales were almost $3 billion in 2015 and are forecast to exceed $4.5 billion by 2022.

Supplements often have a slightly bohemian feel, the muesli and lentils of the medicine cabinet. But don't be fooled: they are big business. According to recent estimates, we collectively buy $128 billion worth of them every year. In the US – the world's biggest market – adults spend an average of $160 each a year.

They might as well flush the money down the toilet. As we've seen here, only a tiny handful do any good. Most are pointless, and some are actively harmful.

If you're thinking about buying a supplement – or already take one or more that you are reluctant to give up – remember that supplement manufacturers do not have to prove in clinical trials that their pills work. As long as there's reasonable expect-ation that the products won't do any harm, they can sell whatever they like. The quid pro quo is that they cannot make explicit health claims about their products, but they don't have to – the 'wellness' industry and its credulous media backers do that for them.

I know how easy it is to get sucked in. I've been known to swallow multivitamins, iron, omega-3 supplements and vitamin C. I don't any more – though I do take vitamin D in the winter and have been advised by a nutritionist to supplement with vitamin B12 as I don't eat meat.

But these are evidence-based choices. Do not swallow the hype. Fools and their money are soon parted.

THE TRUTH ABOUT DRINK (AND DRUGS)

FANCY A DRINK? At some point today the answer will be yes. Humans need to drink to stay alive. If you stopped drinking now you would be dead within a week. It is the only nutrient whose absence is lethal in so short a time.

In our ancestral past there was only one option: water. But while this is the only fluid we actually need to drink to stay alive, modern life has endowed us with a much wider range of options: juice, soda, tea, coffee, sports drinks, smoothies and all kinds of health drinks. All of these liquids play an important role in diet and health but are often overlooked and widely misunderstood.

Many people also drink alcohol. There are even more myths and misunderstandings around this powerful little molecule. And while we're on the topic of dangerous psychoactive substances, we will also be taking a quick detour into the world of recreational drugs, from smoking to methamphetamines.

THE TRUTH ABOUT STAYING HYDRATED

How much you should drink is a surprisingly contentious subject. The average adult drinks about 1.7 litres of fluid a day: water, tea, coffee, soda, milk, fruit juice and more. However, it is common to hear that you need eight glasses of water a day – about 2 litres – even if you don't feel thirsty.

In 2002, physiologist Heinz Valtin of Dartmouth Medical School in New Hampshire tried to track down the source of this advice. The closest he came was a 1974 book that casually advised six to eight drinks a day. As for its scientific validity, Valtin found none.[1]

If anything the number is too low. You lose water all the time, in urine, sweat and water vapour in your breath, and it must be replaced. The US Institute of Medicine sets the general replacement level at approximately 2.7 litres of water a day for women and 3.7 litres for men. Set against that, eight glasses is low.

However, the Institute also points out that all drinks count. Tea, coffee, juice, smoothies, sodas and even alcoholic drinks all contribute. The water in food – especially fruit and vegetables, which are mostly water – also tops up your tank.

In fact, some 20 to 30 per cent of an average person's fluid intake comes from food. Some foods are obviously very watery – watermelon, for example. But some very non-watery foods are surprisingly hydrating. Chicken and salmon are around 60 per cent water, beef 50 per cent and bread 40. A pizza is about 40 per cent too. Even some of the driest foods imaginable – dry roasted peanuts, crackers, cereals and pretzels – are up to 10 per cent water. The only totally waterless foods are pure oils and table sugar.

Another US body, the Food and Nutrition Board of the National Academies of Science, Engineering and Medicine, advises: 'The vast majority of healthy people adequately meet their daily hydration needs by letting thirst be their guide'. The only exception is some elderly people whose feedback mechanisms go awry, meaning they can become dehydrated without thirst.

Generally, there is little to gain by doing more than just quenching your thirst. Contrary to popular belief, water doesn't remove toxins from the skin, visibly improve your complexion or cure constipation. There is some support for the idea that drinking cold water makes you burn calories, and water with a meal does reduce overall calorie intake, perhaps because it helps fill you up or because it means you're less likely to consume a sugary drink as well. But the overall influence of water on weight is far from clear.

There is a sliver of evidence that being adequately hydrated can protect against health problems including colorectal and bladder cancer, heart disease, hypertension, urinary tract infections and kidney stones. Good hydration makes it easier for the kidneys to extract waste, reducing wear and tear on them. Dehydration headaches do exist and water can cure them (although there are hundreds of other reasons why your head might ache), and drinking lots when you have a cold may loosen mucus, easing the symptoms.

Water may not be a cure-all, but the downsides of overdoing it are mild. Besides rare deaths through over-hydration among marathon runners and recreational drug users (especially of MDMA, aka ecstasy), the worst of it is that many people who regularly push the fluids too hard appear to be mildly hyponatremic – they have too little sodium in their blood.[2] This is not a major problem, but has been associated with mild cognitive impairment and an increased risk of falling in older people.

Overall, though, there are few negative effects of water intake and the evidence for positive effects is quite clear.[3]

Perhaps the most implausible claim of all has the strongest support: water can improve focus, at least among children. Several studies have found that having children aged seven to nine drink water improves their attention and, in some cases, recall. Perhaps children of this age are more prone to dehydration, which can cause a decline in alertness, concentration and working memory.

THE TRUTH ABOUT TAP WATER VERSUS MINERAL WATER

For some, tap water is too clean, laced with compounds containing chlorine used to sterilise it. For others it's not clean enough, teeming with nasty pathogens and traces of chemicals. Then there's the fluoride often added to it for dental health: decried as a Communist plot in 1950s America, it remains controversial in some quarters today.

Whether for those reasons or simple taste, many people prefer to buy bottled water. Either way, that could be a waste of money, in most parts of the West at least. Around 25 per cent of bottled water sold in the US is simply tap water from municipal sources. A large proportion of bottled water is chlorinated just like tap water – for good reason. Water chlorination is impressively effective at preventing serious diseases such as dysentery, cholera and typhoid. Evidence that chlorination can produce carcinogenic by-products or compounds that reduce male fertility is 'inadequate', according to the WHO, and the risk is extremely small compared with that from poorly sterilised water.

Tap water does contain traces of pharmaceuticals, toiletries and cosmetics, but the US Environmental Protection Agency says: 'There are no known human health effects from such low-level exposures in drinking water.' As for fluoridation, there is no evidence that this causes any health problems except where accidents lead to over-fluoridation, which can cause vomiting and diarrhoea.

There is stronger evidence that the minerals in some bottled waters, especially sodium, can be harmful. And while the health benefits of mineral-rich waters have long been touted, the enormous variation between brands makes this impossible to test.

Sparkling water is sometimes said to erode your teeth. This is a plausible claim as the bubbles are carbon dioxide, which dissolves in water to become carbonic acid. But the acidity levels are nowhere near enough to erode tooth enamel, so stop worrying about that.

As for the difference in taste, that is a subjective matter. But if it is an issue, just chill your tap water: it makes bad flavours much less noticeable.

THE TRUTH ABOUT SOFT DRINKS

Sugary drinks rot your teeth, and the more you drink, the more teeth will rot.[4] Fizzy pop is generally assumed to be the worst. That is not because of dissolved carbon dioxide – as with sparkling water, the carbonic acid isn't bad for your teeth – but because of the combination of sugar and common flavourings such as phosphoric acid.

Their high sugar content means squashes and sodas deliver a huge calorie hit without filling you up – delivering what are known as 'empty calories'. One standard can of a non-diet soda provides more free or added sugar than the WHO considers healthy. It 'strongly recommends' not exceeding fifty grams a day and advises no more than twenty-five. A 330-millilitre can of classic Coca-Cola contains thirty-five grams. That piles in excess energy that we store as fat. Those who regularly consume sugary drinks are more likely to be overweight, regardless of income or ethnicity, and consuming a can of sweetened fizz or the equivalent a day increases the risk of type 2 diabetes by a quarter.[5] Overall, this form of liquid sustenance has little to recommend it.

You might think sports drinks, another type of soft drink, are healthier. After all, their main claim is that they improve athletic performance and recovery by replacing fluid, energy and electrolytes – sodium, potassium and chloride – sweated out during exercise. A review published in 2000 concluded that sports drinks probably do improve performance compared with drinking water. In 2006 the European Food Safety Authority agreed.

But most sports drinks also come with a stonking sugar hit, and more recent studies have questioned earlier conclusions. An analysis published in the *British Medical Journal* found a 'striking lack of evidence' for any claim related to sports drinks.[6] They may help elite athletes, but are unlikely to do anything for ordinary people.

But there is some evidence that a certain drink can help you recover from an intense workout: low-fat chocolate milk.[7] Its four to one mixture of carbohydrates and protein appears to be ideal for muscle recovery after a workout, and it is cheaper than most alternatives.

Zero sugar

So, if the main problem with sugary drinks is sugar, eliminate that and you eliminate the problem, right?

Not so fast. Some studies indicate that diet sodas help with weight loss, but others find a seemingly paradoxical association with weight gain. Mice consuming artificial sweeteners can even develop glucose intolerance, a precursor to type 2 diabetes.

It is tricky to pin down cause and effect in human studies: people who are already overweight may be consuming diet drinks in an effort to lose weight, skewing the stats. And the animal studies have been criticised as unrealistic, with mice or rats in some experiments consuming quantities of sweeteners equivalent to us gobbling a few hundred sachets a day.

But there are plenty of reasons why low-calorie sweeteners might not always have their intended effect. One is psychology: you had a diet soda this afternoon, so you can have an ice cream this evening. Alternatively, it could be that the intenseness of the artificial stuff, which can be 200 times as sweet as sugar, causes sugar cravings or leads to a general preference for sweetness. Or perhaps sweeteners disrupt our gut bacteria, or our normal hormonal response to sugar intake, and as a result the body doesn't respond as well when real sugar is consumed and weight gain follows.

The scientific consensus is that choosing diet drinks over normal sugary drinks can contribute to weight loss.[8] But the uncertainty should keep your finger hovering over the ring pull.

THE TRUTH ABOUT FRUIT JUICE

Pure fruit juice feels like it ought to be a healthy drink. It's 100 per cent fruit, after all, and contains good stuff that fizzy drinks don't, such as vitamins, minerals and antioxidants. The UK National Health Service says one small glass (150 millilitres, about a quarter of a British pint) of pure fruit juice can be counted towards your five a day.

But only one; any more don't count. That is because fruit juice is missing a lot of the goodness in whole fruit. The juice of one orange contains 0.4 grams of fibre, compared with 1.7 grams in an actual orange. And it is as sugary as sweetened drinks: that 150-millilitre glass contains nearly 10 grams of sugar. The WHO recommends that the natural sugar in fruit juice should be lumped together with that added to food and sweetened drinks as free sugar, and advises strict limits on how much you should consume. Orange juice and Coca-Cola contain roughly the same amount, and some juices even more – red grape juice is about half as sugary again, serving up about 25 grams per 150-millilitre glass. That suggests pure fruit juices should carry the same health warnings as added-sugar drinks.

In truth, we don't know whether fruit juices are better or worse for you than soda. Other lifestyle factors such as income, diet, smoking and exercise that may differ between habitual juice drinkers and habitual soda drinkers make it hard to draw water-tight conclusions. But as a rule of thumb, the first 150 millilitres aside, juice and soda are as bad as one another.

THE TRUTH ABOUT COFFEE AND TEA

Tea and/or coffee are a part of many people's daily routine. In 2016, a study found that the world's top-consuming tea nations were the UK, Ireland and Turkey, with the Turkish getting through more than three kilograms of tea per person each year – enough to fill 1,200 tea bags. Another study in 2016 placed Finland at the top of the list of the world's biggest coffee consumers, drinking twelve and a half kilograms of coffee per person every year – enough for 1,800 cups. How much do we really know about tea and coffee, drunk by such a large proportion of the world's population?

Coffee

Coffee often gets a bad rap. Critics point out that it is full of caffeine, which is addictive and can make you irritable, twitchy and unable to sleep. Excessive consumption has been linked to heart disease and cancer. And although coffee increases alertness and focus, the effects are quite short-lived and heavy users quickly become tolerant. People who regularly drink coffee are no more alert on average than those who don't. For caffeine addicts, a morning coffee merely reverses the symptoms of caffeine withdrawal, bringing them back to normal.

For these reasons, many people avoid or keep a lid on their coffee intake. But that may be misguided. If good health is your overall goal then coffee looks a pretty good option. A recent study of almost half a million Britons, tracking their coffee intake over ten years, found 'inverse associations for coffee drinking with mortality' for the consumption of up to eight cups a day.[9] In other words, the more coffee people drink the less likely they are to drop dead of any cause. This finding chimes with others from North America, Asia and other parts of Europe.

Exactly why is not known: it may be that coffee itself is good for you, or that drinking it is associated with a generally healthy lifestyle. Caffeine itself doesn't seem to be the magic ingredient as the association still holds up for people who drink decaf.

One possibility is that coffee contains high levels of compounds called chlorogenic acids, known to slow the body's absorption of glucose, which may lower the risk of type 2 diabetes. But whatever the explanation the results give little reason to demonise coffee.

There is a type of coffee to avoid, though. Two oily compounds in coffee, cafestol and kahweol, do seem to increase the bad cholesterol that clogs blood vessels. But most of the coffee we drink, including instant, doesn't contain much of either. Espresso machines almost entirely filter them out and French presses don't do a bad job either. The type to avoid is the boiled, unfiltered coffee popular in Turkey, Norway and Sweden.

Turkish-style coffee has also been associated with cancer. But studies of consumption of other types of coffee typically find no correlation or a mildly beneficial effect, except among people who drink forty (!) or more cups a day. The WHO recently changed its stance on coffee from 'possibly carcinogenic' to 'no conclusive evidence'. The sole caveat was that any hot drink – above 70°C – increases the risk of oesophageal cancer. So enjoy coffee, but do your digestive tract a favour and let it cool. And be aware that any milk, cream and/or sugar you add must be taken into account when considering your overall diet.

It seems to be good news all round for coffee. Research funded by the British Heart Foundation found that there was no increased risk of artery disease among people drinking up to twenty-five cups a day. The effects of caffeine on your bladder have also been overblown. It is mildly diuretic but the amount of water you drink it with will almost certainly exceed what you lose. Again, let thirst be your guide.

And if addiction is a worry, be reassured that caffeine is surprisingly easy to kick. Simply cutting intake gradually over a few weeks is effective.

Tea

Unlike coffee junkies, tea drinkers are often seen as being infused with health. Their beverage of choice is full of healthy plant compounds called polyphenols, which boast antioxidant and anti-cancer effects, at least when added to cells grown in a Petri dish. Green tea – which is less heavily processed than black tea – is especially high in polyphenols. Herbal and fruit teas also contain them, though the quantities vary a lot from variety to variety.

Tea (meaning actual tea, an infusion of the leaves and buds of the *Camellia sinensis* plant rather than fruit or herbal teas) also contains caffeine, but as with coffee that doesn't seem to be a problem per se. Tea leaves are actually higher in caffeine than coffee beans but a typical cup of tea contains less caffeine than a typical cup of coffee because tea is brewed weaker.

But just as coffee isn't as bad as it is made out to be, tea isn't as good. Although some studies have found that drinking green tea (and, to a lesser extent, black tea) lowers the risk of breast, gut and lung cancers, a review of fifty-one studies involving a total of 1.6 million people concluded that the evidence was patchy and contradictory.[10]

Like coffee, hot tea also causes oesophageal cancer. Again, milk and sugar must be taken into account. Still, a nice cuppa is unlikely to do you much harm.

Tea is good for your teeth, too, unless you drink insane quantities. One woman did lose all her teeth at forty-seven due to a fluoride overdose from tea, but she had been drinking more than a hundred cups daily for seventeen years. For a normal intake, tea's fluoride content and antibacterial properties protect our teeth. A study of green tea as a mouthwash found that it killed just as many bacteria as a standard chlorhexidine-based version, and would probably work out cheaper.[11] Black tea similarly fights cavities and stimulates the mouth's own antibacterial enzymes.

THE TRUTH ABOUT HEALTH DRINKS

Coconut water, wheatgrass smoothies, vinegar, beetroot juice, even urine – the list of trendy cure-all potions is growing. But how many of them actually work?

Coconut water
When it comes to hydration, what could be better than water? If you believe the hype, the answer is coconut water. This cloudy

liquid tapped from young, green coconuts is supposedly 'hyper-hydrating' and has been dubbed nature's sports drink. But studies comparing it with ordinary water have found no difference in terms of how well it hydrated volunteers after vigorous exercise.

The claims for coconut water rest on the idea that its higher potassium levels enhance water absorption. That doesn't stand up: according to another recent study, neither coconut water nor a potassium-rich sports drink scored higher on fluid retention than water. Besides, you shouldn't have any problem absorbing water so long as your diet contains adequate amounts of salt.

There are other supposedly 'hyper-hydrating' waters out there including watermelon water and birch sap water. As yet there is no scientific evidence that they are any better than water.

Beetroot juice

Beetroot juice is rich in nitrates that can relax blood vessels and improve blood circulation and there is some scientific support to the idea that it is good for you.[12] But drink it in moderation: its sugar content is on a par with orange and other common fruit juices. Too much nitrate has also been tentatively linked with an increased risk of stomach cancer.

Wheatgrass

Wheatgrass contains a smorgasbord of vitamins and minerals, as well as chlorophyll, claimed by some to boost the production of red blood cells. But studies show it is unlikely to benefit you much more than eating green vegetables such as broccoli and spinach.

Kefir

A fermented milk drink akin to yoghurt, kefir is prized for its supposed beneficial effects on microbes in the gut. Studies in mice suggest there might be a link – although it is too early to say whether there is an effect in humans, or how big it is.

Urine

Here's a drink for the very adventurous: a nice cup of wee.

Drinking urine (usually your own, but there's no accounting for taste) is often portrayed as the last resort for somebody lost in the wilderness far from water. But drinking it for health reasons has a long history and has recently come back into vogue. Fans of urophagia claim it can cure all sorts of ills: acne, anaemia, allergies, obesity and various cancers. The rationale is that urine contains vitamins, minerals, proteins, enzymes, hormones, antibodies and amino acids your body has discarded.

But there's a reason your body discarded them. It either doesn't need them, or they are toxic. Unsurprisingly, there's no scientific evidence that drinking urine is good for you. It won't even hydrate you – those waste products will have to be peed out all over again. Advice from the real survival experts is unequivocal. When it comes to preserving precious bodily fluids, the US army's 1999 survival field manual ranks urine alongside blood, seawater and fish juices in its 'Do Not Drink' category.

So drinking urine is as pointless as it is disgusting, but at least it is harmless, right? Maybe not. One component of urine rarely mentioned by those who promote drinking it is phosphorus, a possible cardiac toxin. The myth that urine is sterile is just that – a myth. Drinking it could bring you down with all sorts of bacterial diseases.

Just to reiterate, in case you're still not convinced: *do not drink your own urine – or anybody else's.*

Vinegar

Speaking of repulsive liquids that most people wouldn't touch with a bargepole, let's talk about vinegar. Cleopatra supposedly dissolved pearls in vinegar to make Mark Antony a love potion. Nowadays it is more likely to be drunk as an appetite suppressant and weight-loss aid.

Numerous small studies have been done to test this but they vary so much in design – from the amount and type of vinegar to the health of the subjects – and outcome that it is impossible to draw broad conclusions.[13] If vinegar does suppress appetite, that may be down to the fact that drinking it is nauseating.

The good news is that if you can keep it down, a bit of vinegar might add some useful nutrients to your diet. It contains a teeming collection of amino acids and polyphenolic compounds. But so do hundreds of foods, and any benefits must be set against the damaging effects of acetic acid on tooth enamel.

THE TRUTH ABOUT ALCOHOL

What's your poison? For billions of people around the world, the answer is a small organic molecule that is excreted by yeast when it digests sugar. Chemists call it ethanol; ordinary people know it as alcohol. The demon drink.

The health benefit versus health risk of alcohol is one of the most hotly debated topics in nutrition science. A lot of people believe that drinking in moderation is good for them, especially if that drink is red wine. With more than a hundred published studies finding a link between moderate alcohol consumption and a decreased risk of heart attack or stroke, it is tempting to raise a glass to the stuff.

But only one, two at the most. Moderate alcohol consumption means one or two drinks a day, with a drink defined as ten grams of alcohol (1.25 UK units). A study that analysed data from more than 300,000 people over twelve years found none of these cardiovascular benefits for those who consumed more than four drinks a day. And there are good reasons to believe that even this restrained level isn't actually good for you.

A confounding factor in most such studies is that people who drink in moderation also tend to have other lifestyle habits that lower heart disease risk. They exercise more regularly, have a healthier weight, sleep better and are more affluent. These factors may mask the harms of drinking alcohol. The results may be further confounded by the fact that many studies don't differentiate between those who have never drunk alcohol and those who have given up drinking, perhaps because they have already done themselves damage through too much booze. Their health problems may mask the benefits of abstention.

Despite many studies showing a small benefit of moderate drinking, there is no scientific consensus. There are people who are passionate that there are health benefits, others equally passionate that it is all down to errors in the data.

Official advice now errs on the side of abstention. A major WHO study published in 2018 concluded that there is no such thing as a safe level of drinking, let alone a beneficial one.[14] Alcohol consumption is bad for you, period. Far from decreasing the risk of cardiovascular disease, it increases it. It also causes liver disease and at least seven types of cancer: it is classed as a group 1 carcinogen, meaning it definitely causes cancer in humans. It also weakens the immune system, as well as increasing the risk of death from accidents and violence. And if you're tempted to dismiss this as just one piece of advice among many, consider the fact that the study tracked millions of people in 195 countries and territories over sixteen years.

To reflect this growing understanding of alcohol's true health impact, the UK government recently tightened its guidance on limits, recommending no more than fourteen units a week – the equivalent of six pints of beer – for both men and women, spread evenly over three days or more. That is a big change in a short time. In 1979, it was fifty-six units a week for men. These lower limits are deliberately stringent. Two and a half million people regularly drink more than fourteen units in a single sitting.

Many countries allow men to drink more than women, but the UK advice is based on assessments of overall harm. Alcohol does affect men and women differently. Women have a higher blood alcohol concentration after drinking the same amount, but metabolise it faster. Women are also at higher risk for some alcohol-related cancers. But men are more frequently heavy drinkers and tend to engage in more risky behaviour. Overall these factors balance out.

One thing national guidelines can't do is account for how risk varies between individuals. Lots of things determine how bad alcohol is for each of us, from social circumstances to mental health issues. Even our genes may play a part: studies in mice show associations between particular genes and propensity to drink alcohol.

How you drink also matters. Studies of consumption in France and Russia show that drinking some wine with meals every day is safer than drinking the same amount of vodka in a single binge. I'll let you guess which country followed which pattern.

That finding may also have something to do with the purported health benefits of wine, especially red. It is full of resveratrol, a plant chemical thought to protect against the effects of stress and a poor diet. It has been linked to a reduction in age-related conditions such as arthritis, macular degeneration and dementia. While many of these studies have been in animals, human research has suggested that resveratrol in pill form can slow the progress of Alzheimer's. One catch: to consume the same amount of resveratrol, you would have to glug 1,000 bottles of red wine a day.

And even if this molecule is good for you, it comes from grapes. So in the unlikely event that you are drinking wine solely for the polyphenols, it would be much healthier to eat red grapes instead.

There's another get-out clause that you could try: what about the relaxation and social lubrication that alcohol delivers? Surely that is a good stress buster, and we know that stress is a stealth killer (for more on why stress is so bad, see page 252). There may be something in that, but it is very hard to quantify. If it were a major factor it would be picked up by epidemiological studies of the mortality of light and moderate drinkers versus non-drinkers. We've already dealt with that.

How risky exactly is drinking? The WHO study calculated that consuming ten grams of alcohol a day increases your risk of developing one alcohol-related disease by 0.5 per cent.[15] That is a small increase in risk, but it rises steadily the more you drink.

For example, another large multinational study found that drinking more than a hundred grams of alcohol every week – just over a bottle of wine or five pints of beer – increases the risk of death.[16] A forty-year-old who regularly drinks between 200 and 350 grams of alcohol per week lowers their life expectancy by around one to two years.

That is sobering. But nothing in life is risk-free and even the authors of the study consider that figure 'neither practically nor statistically significant'. It means that 25,000 people would need to drink ten extra grams of alcohol a day for one additional person to develop a drink-related disease each year. You might well decide that the risk is worth it, just as most of us happily get into cars or eat salty food.

Clearly, though, many people are taking too much of a risk. The WHO estimates that alcohol kills three million worldwide every year, and is responsible for 12 per cent of deaths in males aged fifteen to forty-nine.

If you're concerned about drink then one well-known strategy for cutting down is to do a dry January. The idea is that by quitting booze for a whole month, you give your body breathing space to recover from past excesses and reset your relationship with alcohol.

In the short term at least, a dry month appears to have health benefits. A study of moderate to heavy drinkers who did it found reductions in blood pressure, cholesterol and two growth factors associated with the development of cancers. It also found a significant drop in insulin resistance, suggesting their risk of developing type 2 diabetes went down.[17]

But whether dry January changes long-term behaviour is less clear. Only one study has directly assessed this.[18] It found that people who completed the challenge claimed to be consuming less alcohol six months on. However, 36 per cent of participants failed to complete the challenge, and for this group, the reduction in alcohol consumption at six months was smaller.

Some alcohol specialists are wary of dry January because they fear it gives people the impression that they can drink without limits for the rest of the year. Much better, they say, would be to go 'damp' rather than dry – commit to having at least two consecutive alcohol-free days a week, every week. Experts say this reduces the impact of alcohol on your liver and breaks the habit of drinking every day.

Another way of looking at it is through the medium of the 'microlife'. This is a useful way of getting your head around the impact of lifestyle choices that almost certainly won't kill you today but still chip away at your health over the long term. A microlife is a millionth of a life, and equates to about half an hour.

Microlives are designed to overcome the psychological block we have when considering the long-term consequences of our actions. This is called future discounting, which means that we tend to value short-term gain (a pint) over long-term pain (alcohol-related disease).

Microlives remind you that the pain is happening now. For alcohol, each unit knocks half a microlife off your total. To put that into perspective, smoking one cigarette also claims half a microlife, as does sitting in front of the TV for an hour. Twenty minutes of physical activity adds a microlife. So if you do drink, think about exercising too.

In fact, don't just think about it. Do it. A recent analysis of data from 36,370 people aged forty and over showed that in those who did little to no exercise there was a direct relationship between alcohol intake and cancer. But in those who drank moderately and did at least seven and a half hours of activity a week the risk of death from cancer and heart disease was decreased.[19]

THE TRUTH ABOUT HANGOVERS

If you've ever had a hangover, you can at least comfort yourself that you're continuing a long and ignoble historical tradition. Descriptions of hangovers are known from ancient Egyptian and Greek texts and even the Old Testament. About half of people give themselves at least one each year.

And yet the 'hangover syndrome', as doctors know it, remains quite mysterious. There are forty-seven recognised symptoms, but many have an unknown cause and preventing or curing them remains beyond the ken of medical science. According to a 2011 survey of 1,410 hung-over Dutch students, this is the order of frequency of hangover symptoms: fatigue, thirst, drowsiness, sleepiness, headache, dry mouth, nausea, weakness, reduced alertness, concentration problems, apathy, increased reaction time, reduced appetite, clumsiness, agitation, vertigo, memory problems, gastrointestinal complaints, dizziness, stomach pain, tremor, problems with balance, restlessness, shivering, sweating, disorientation, audio sensitivity, photosensitivity, blunted emotions, muscle pain, loss of taste, regret, confusion, guilt, gastritis, impulsivity, hot/cold flushes, vomiting, pounding heart, depression, palpitations, tinnitus, nystagmus (uncontrolled eye movement), anger, respiratory problems, anxiety, suicidal thoughts.[20]

The main reason why hangovers make you feel rough is – you guessed it – the effects of drinking too much ethanol, that magical and demonic little molecule that gets us drunk. Many classic hangover symptoms, including headache, nausea, sweats, tiredness, apathy and self-loathing, have all been directly linked to ethanol.

But ethanol isn't the whole story. Fatigue could also be down to staying up too late to find the time to get all that drinking done. Or it could be that alcohol – and maybe an end-of-the-evening takeaway – have a detrimental effect on sleep quality.

Other symptoms, like dehydration, are side effects. Your dry mouth and pounding head probably result from alcohol's effect on your bladder. It suppresses the antidiuretic hormone vasopressin, which makes you urinate more.

As for the other symptoms, nobody is really sure. The morning-after onset of a hangover means that the prime suspects are the breakdown products of ethanol – especially acetaldehyde, a toxic compound produced in your liver as the first step of ethanol metabolism.

You can take solace from the fact that once the acetaldehyde has been cleared from your system you will feel a lot better. But your organs aren't so lucky. They will be feeling the effects of acetaldehyde for much longer.

For such a simple molecule, acetaldehyde is surprisingly toxic. It reacts with proteins to form stable compounds called adducts, which cause irreversible damage by messing up the protein's structure and function. After a drinking session, adducts are formed in the liver, muscles, heart, brain and gastrointestinal tract. Acetaldehyde also attacks DNA; the International Agency for Research on Cancer classes acetaldehyde as a group 1 carcinogen, meaning it definitely causes cancer in humans.

The body has a battery of enzymes that break down acetaldehyde to acetate, which is harmless and used for energy by the muscles. They do a great job, rapidly eliminating over 99 per cent of the acetaldehyde. But minute amounts escape into the bloodstream, where it goes on a bender.

The average liver can process about seven grams of ethanol an hour – though in heavy drinkers that figure can rise to ten grams – meaning that it takes about twelve hours to eliminate all the ethanol in a bottle of wine. That's twelve hours of continuous exposure to a trickle of poisonous, carcinogenic acetaldehyde.

There may be other contributors to your hangover. Various studies have looked into the role of 'congeners' – other organic chemicals produced during fermentation that give each drink its distinctive taste and aroma. These include acetaldehyde and other types of alcohol, including toxic methanol, collectively known as 'fusel alcohols'. *Fusel* is a German word that roughly translates as 'gut rot'.

As a rule of thumb the darker a drink the more congeners it has in it, and the more likely it is to give you a hangover. And given that different drinks contain different congeners, one sure-fire way to a bad head is to mix your drinks. Red wine and whisky is a particularly potent congener cocktail.

Some research suggests that congeners inhibit the breakdown of ethanol so at least delay the onset of hangover. But this is a difficult balancing act to strike. And in any case by the time you're on to the dark spirits, it is probably too late.

As for the folk wisdom about beer before wine making you feel fine, and wine before beer making you feel queer – or is it the other way round – that is simply not true. A rigorous test of the hangover potential of both found no difference.[21]

Rather obviously, the most important factor in determining hangover severity is how much you drank. But this rule doesn't always hold. As any experienced drinker knows, it is possible to inflict a horrible hangover on yourself without drinking heavily and also to get off scot(ch)-free after a skinful. This is may be down to other factors, such as how fast and exactly what you drank, whether you were already dehydrated or tired, what and when you ate, how you slept, even your mood.

This is the straw that breaks the back of an already flimsy body of hangover research. There are too many variables in how we actually go about getting hungover to establish reliable correlations. Asking people the next morning what they drank and in what order – plus other possibly relevant variables such as whether they drank water, when and what they ate, whether they smoked and so on – is asking for trouble. As with food diaries, people forget and lie. Given that alcohol is a memory suppressant, there's too much margin for error.

Hangovers supposedly get worse with age, but there has never been a study with a statistically large enough sample to test this.

The best way to cure a hangover is to avoid one in the first place. But if you do wake up to the morning after the night before, there are a few things you can do.

Drinking water will alleviate dehydration and painkillers ease the headache. Caffeine will increase alertness. If you don't have to get up, stay in bed with the curtains closed. Darkness has been shown to speed recovery, at least in mice.

Some people swear by vigorous exercise, which at least won't do you any harm and may speed up the metabolism of residual ethanol and acetaldehyde. But there's no solid evidence that it works.

There is no evidence, either, that depleted levels of electrolytes such as sodium, potassium and magnesium are correlated with hangover severity. So an isotonic drink is likely to do little more than slake your thirst; anything over that is the placebo effect. Ditto all the other 'proven' cures – greasy food, raw eggs, bananas, fresh air, ice packs, saunas, vitamins. The placebo effect can be powerful, however – even when you know it's a placebo (for more on the placebo effect, see page 224). So if it works for you, go for it.

But be very leery of the dedicated drunk's remedy, hair of the dog. It may make you feel better temporarily but in reality is only delaying the inevitable.

It is unlikely there will ever be a really reliable cure. Hangovers are complex, multifactorial beasts. And doing research on them is difficult.

Researchers have three options, all of them unpalatable. First, there's the notoriously unreliable self-report. Second, animal experiments, which are suspect ethically and of questionable validity, given the human factors that contribute to hangovers. Third, laboratory experiments with human volunteers under controlled conditions. These again are only a vague simulacrum of real drinking behaviour and also introduce an observer effect: people do not behave normally if they know they are being watched.

Maybe it is better all round that we can't cure hangovers. Research suggests that fear of the morning after helps to keep a lid on people's drinking. A hangover is nature's way of telling you not to do it again. So, being human, you'll go out and do it again.

THE TRUTH ABOUT SMOKING

Smoking is one of the worst things you can do for your health. For a habitual smoker, each cigarette shaves about fifteen minutes off their lifespan. But what about social smokers, who might have an odd puff outside the pub, or passive smokers who are exposed to other people's second-hand smoke?

According to the British Heart Foundation, there are 1.1 million occasional smokers in the UK. That includes people who smoke but not every day, or smoke an average of less than one cigarette a day. Can the health effects of such a habit really be that bad?

Some of the risks of occasional smoking do pale in comparison to those from a twenty-a-day habit. The risk of lung cancer, for example, increases linearly the more you smoke.

But there is no such thing as no risk. Every cigarette you smoke – or every lungful of second-hand smoke you breathe in – increases your chance of getting lung cancer and other lung diseases. Fine particulate matter in the smoke causes damage to the lining of the lungs, the cumulative effects of which can be a serious breathing disorder.

And social smokers who think the harms can be compensated for by exercise, diet or antioxidant supplements are kidding themselves. There is no way to protect your lungs from this damage or fully reverse it once it is done. Even stopping smoking is not going to restore your lungs to a pristine state.

The biggest immediate risk for social smokers is cardiovascular disease and heart attack. Unlike the risk of cancer or respiratory disease, which increase linearly with every cigarette smoked, the risk of heart disease is non-linear, with the highest jump coming after the first cigarette and increasing more slowly thereafter.

If your relative risk at zero smoke exposure is 1.0, one cigarette increases it to 1.4. Going from one to five cigarettes only raises it to 1.5, and it rises by another 0.1 for every five cigarettes thereafter. In other words, to double your heart attack risk you need to smoke thirty a day. Passive smoking raises it to 1.3.

But be aware that this is relative risk; in other words, the increase in risk of having a heart attack over your baseline risk, which varies from person to person. So if you have a very low risk of a heart attack because of being fit and having a normal

BMI, low blood pressure and cholesterol, a crafty fag won't raise it very much. But is it worth it?

Quitting smoking altogether will undo some of the risk damage. Within a year your relative risk of smoking-related heart disease will be halved.[22] Quitting could also claw back some years lost to your lifespan. Among people who die of cardio-vascular disease, smokers do so on average five and a half years earlier than non-smokers. For ex-smokers, the gap goes down to two and a half years – even if they quit as late as their sixties.

You may have heard of the 'smoker's paradox', whereby smokers admitted to hospital after a heart attack have a better chance of survival than lifetime non-smokers. This is a bit like the smoking equivalent of the claim that a little bit of alcohol does you good, and is equally untrue.

The protective effect could be because of some of the bad effects of smoking, such as over-activation of the body's inflammatory response, or causing platelets in the blood to form clots; it also applies to traumatic injury. However, this benefit is gigantically overshadowed by its harms and is a poor excuse for carrying on smoking.

That is another good reason to quit. But quitting is often easier said than done, and being an occasional smoker does not make it much easier. Even people who smoke less than one cigarette a day struggle to give up completely when they want to, with 65 per cent relapsing within six months of their attempt to quit.

According to Public Health England, the most effective way to quit smoking is to use your local stop-smoking service, which instantly gives you a support network of fellow quitters and health professionals, and can also offer behavioural therapy and pharmaceuticals. Even so, only 16 per cent of people who go down this road succeed, which means being smoke-free a year after joining. The least effective method is willpower alone. E-cigarettes are controversial but most health professionals believe that they can help people quit and are less damaging than cigarettes for people who can't or won't give up nicotine.

A newer nicotine delivery system called heated tobacco products (HTPs) or heat-not-burn makes all sorts of claims about delivering vastly fewer harmful compounds than a cigarette while being more similar to one than a vape. Tobacco companies are increasingly urging smokers who cannot or will not quit to switch to HTP; one product is even called IQOS, which stands for I Quit Ordinary Smoking. But scientists say that, even if the claims about harmful compounds in the vapour are true, it does not follow that they are less damaging to the lungs. It is too early to say what the long-term health effects of vaping and HTPs are.

If you really want to quit smoking, there are a few small behavioural changes you can make to increase your risk of success. These include making a plan and sticking to it, having a strategy to work through cravings, such as going for a brisk walk, and watching what you eat. For unknown reasons some foods, especially meat, make an after-dinner cigarette more satisfying. Others, including cheese, vegetables and fruit, make cigarettes taste horrible. And if you fall off the tobacco wagon, don't beat yourself up. Having a positive mindset can be the difference between success and failure.

THE TRUTH ABOUT RECREATIONAL DRUGS

One of the most useful ways to look at alcohol and smoking is in comparison with other recreational drugs. Perhaps the best guide to this comes from the UK's Independent Scientific

Committee on Drugs (ISCD), which analysed twenty drugs on sixteen measures of harmfulness.[23] These included short- and long-term health problems, addiction and accidents.

It found the most harmful illicit drug to be heroin, with an overall rating of fifty-five out of a hundred, with crack cocaine on fifty-four. LSD and magic mushrooms are among the least harmful, and also carry the lowest risk of dependence. The world's most popular illicit drug, cannabis, scores twenty. Tobacco is the sixth most harmful on twenty-six, just below cocaine and just ahead of amphetamines. But what is the most damaging drug of all? Alcohol, with an overall harm score of seventy-two.

That may tempt some people to look for an alternative to alcohol. If so, you're spoiled for choice. Besides the traditional menu of cannabis, cocaine, speed, ecstasy, LSD and mushrooms, there's a growing 'long tail' of novel psychoactive substances, meaning that psychonauts can explore the outer reaches of their consciousness in ever more creative ways.

But beware. There's a reason why drugs have a reputation for being bad for you: they are. Drug harms fall into two broad categories: those that affect you, and those that affect others. The personal ones include death, health problems (including mental health), accidents, addiction, relationship breakdown and legal trouble. Harms to other people include violence, financial problems, crime and environmental damage – both at home and where the drugs are produced.

One rule of thumb is that risks become more serious with repeated use. Take addiction, for example. According to the US National Institute on Drug Abuse, it can take only 'a few' uses of a drug to become addicted to it, although the potential for addiction varies between drugs and people. Putting firm numbers on this is difficult, but a study published in 2005 found that among a large cohort of people who tried cocaine for the first time, more than one in twenty were dependent on it two years later.[24]

Mixing drugs amplifies the risks. Taking cocaine with amphetamines or ecstasy, for example, raises the risk of acute toxicity over and above the sum of their parts. This also extends to nicotine.

And, of course, most of these drugs are illegal in many places. As well as the potential for falling foul of the law, users often can't be sure what they are taking. Some nightclubs offer a testing service to analyse the contents of party pills, but on the whole the only 'guarantee' is the word of the drug dealer.

When it comes to the benefit side of the equation, the picture is even less clear. Nobody has yet done an analysis taking into account the pleasure, fun and adventure that people seek when they take drugs, and whether that can have a long-term impact on health. Researchers studying the use of psychedelics such as LSD and psilocybin as a treatment for depression report that even a single dose can make people more open to experience, less authoritarian, more politically liberal and more connected to nature and other people. But whether these effects have positive health benefits is not known.

All told, at least from the perspective of health and well-being, it is wise to conclude that taking drugs isn't worth the risk.

If you found this chapter thoroughly sobering, you're not alone. I did too. Water is OK but I really like juices and sodas. I probably down far more artificially sweetened drinks than is good for me. I drink a lot of coffee and tea – always without milk – and now worry that I don't let them cool enough to avoid increasing my risk of oesophageal cancer. If I drink water, I go for the sparkling variety. Looks like I need to wake up and smell the coffee.

And then there's *drink* drink. For me, the most sobering moment of the whole book was discovering that there is no such thing as a safe level of alcohol, let alone a beneficial one. I think I knew that already but I clung onto the belief that moderate drinking was at least harmless and maybe even slightly beneficial. I now see that for what it is: a self-serving delusion. As I confessed in the Introduction, I can't remember the last time I drank less than the recommended upper weekly limit of fourteen units. To be brutally honest, I rarely achieve the two alcohol-free days a week. That is something I will now strive to achieve, and use as a platform to go further.

As for those crafty fags, it is time to quit completely. And my drug-taking days are definitely over. Nowadays I seek my highs elsewhere – in the gym, and out on the streets in a pair of running shoes. Which brings us neatly to the next chapter.

THE TRUTH
ABOUT EXERCISE

IF YOU WANT to be healthy, slim and youthful, there's a miracle treatment that you ought to try. It is available to almost everyone whenever they want it, in unlimited quantities. It doesn't cost anything except a bit of time and effort. It can make you feel as good as any drug. In return you get protection against heart attacks, strokes, diabetes, obesity, cancer, Alzheimer's disease and depression. It can even cure some diseases. Overall, it has the potential to prevent more premature deaths than any other single medical intervention, with none of the side effects.

What is the name of this wonder drug? Exercise – physical activity of all sorts, from walking to the shops to running ultra-marathons. We all know that exercise is good for you and laziness is not, but what is becoming increasingly clear is the sheer extent of its benefits and the downsides of missing out.

At this point you're probably feeling one of two things: either a warm glow because you already exercise, or a sinking feeling because you don't. Whichever camp you fall into, don't assume what follows is not for you. Exercise is both easier to do, and harder to do well, than you might think.

THE TRUTH ABOUT WHY
EXERCISE IS GOOD FOR YOU

Exercise has always been seen as the foundation of a healthy life, but it was only quite recently that the true extent of its benefits was established. An important 'aha' moment came from a 1953 study on employees of London's iconic double-decker buses, which at that time had conductors as well as drivers. The drivers sat down all day but conductors spent a lot of time on their feet, often walking up and down stairs. A team of medics discovered that they had half as many heart attacks as their driver colleagues.[1]

The most obvious benefit of exercise is that it makes you physically fit. That is a rather nebulous concept, but all it really means is that your body has undergone biological adaptations making it able to undertake physical activity more efficiently and for longer.

Fitness is technically defined as aerobic fitness, also known as cardiorespiratory or cardiovascular fitness. This boils down to how effective the body is at delivering oxygen to muscle cells.

If you become more active, your body undergoes numerous changes that boost aerobic fitness. Muscle fibres grow and are better supplied with blood vessels. Within muscle cells, the mitochondria, which release energy from glucose, increase in size and number.

In particular, your heart undergoes significant changes. Ultrasound scans show that the heart of an athlete looks quite different from that of a couch potato. The left ventricle – the chamber that does most of the work of pumping blood around the body – is likely to be much larger, and the walls more muscular. And when our hearts pump hard, the stresses imposed by the blood rushing through arteries promotes the production of nitric oxide, a muscle relaxant that keeps blood vessels elastic and helps repair them.

Exercise also reduces chronic inflammation and blunts our physiological response to stress, reducing the risk of heart disease, cancer and diabetes.

Your bones benefit too. Bone is not just an inert scaffold but a living tissue that responds to exercise by becoming stronger and denser, which is beneficial at any age but especially so in later life when bone density declines and the risk of osteoporosis increases.

But 'fitness' isn't just about stronger muscles and bones and an enhanced capacity for exercise. It has so many knock-on benefits for general health that increasing it even a little bit has got to be a worthy goal for everybody.[2]

The most robust evidence comes from the Exercise is Medicine initiative pioneered by the American College of Sports Medicine in Indianapolis. Researchers there have collated studies of people who follow the US government's advice on physical activity. This prescribes 150 minutes per week of moderate-intensity aerobic activity, such as brisk walking, ballroom dancing or gardening; or seventy-five minutes of more vigorous activity such as cycling, running or swimming. The Exercise is Medicine findings show that this weekly dose of exercise reduces the risk of premature death through heart disease by 40 per cent, approximately the same as taking statins.

How physical activity prevents cardiovascular disease is not hard to understand. Exercise gets blood coursing through the circulatory system, which flushes out fatty deposits in the walls of blood vessels that otherwise threaten to become like mini-fatbergs, clogging up blood vessels and causing a heart attack or stroke.

Exercise also helps to reduce levels of the most dangerous fats. Many risks to circulatory health come from very low-density lipoproteins (VLDLs). The bigger the VLDL particles are, the less dangerous they are, and the findings show exercise causes the particles to enlarge by about a quarter.

It is not just cardiovascular disease. The recommended dose of exercise lowers the chances of developing type 2 diabetes by 58 per cent, twice the preventive power of the most widely prescribed anti-diabetes medication, metformin.[3] It can also help to reverse type 2 diabetes and, of course, is part of most successful weight-loss regimes.

Exercise also reduces the risk of cancer, though exactly how is unclear – it may have something to do with reducing obesity, which is a known risk factor for some cancers. Ditto neuro-degenerative diseases. Exercise also has psychological benefits, guarding against depression and anxiety, for example.

Maintaining fitness is particularly important as we age. Older adults who can walk at least 365 metres in six minutes have half the risk of dying in the subsequent decade as those who can't do 290 metres. Resistance exercises such as weight training help to maintain muscle mass, and loss of muscle – called sarcopenia – is now seen as a contributor to many age-related diseases.

THE TRUTH ABOUT LAZINESS

Laziness is not just not good for you, it is positively bad. Unfortunately, we humans are naturally lazy lumps, and we generally have to kick ourselves up our backsides to get them into gear.

Some other animals are more fortunate: fitness comes easily to them. Bears come out of hibernation in good physical condition despite having done no exercise for months, and barnacle geese can embark on a 3,000-kilometre migration with little more preparation than lazing around eating. Exactly what physiological adaptations these creatures possess to make them resilient to becoming unfit is not well understood. But whatever they are, we don't enjoy their benefits. For humans, physical fitness is induced by physical activity.

We have another biological obstacle to overcome. Whereas mice and dogs appear to crave physical exercise, we seem programmed to prefer laziness. That probably goes back to our evolutionary past, when food was scarce and life unpredictable. Most of the time our ancestors would have been forced to forage, hunt or run away from danger; it was exercise or die. Not exercising was never an option, so there was little evolutionary pressure to make us enjoy exercise. If anything, there was pressure in the opposite direction – to avoid needless exercise. In times of plenty and safety it made evolutionary sense to park our backsides and not expend too much energy.

During these good times it was also adaptive to lose muscle mass. Muscle is quite an expensive tissue, requiring about fifteen kilocalories per kilogram to maintain. That mounts up: muscle accounts for about 40 per cent of the average person's body mass, so most of us are spending 20 per cent of our basic energy budget taking care of them.

This laziness instinct was fine on the savannah, when the next foraging/hunting/escaping trip was just around the corner. But in the modern world where highly nutritious food is available almost all the time and running away from danger rarely necessary, it has become our default position.

This is a very recent development. Even after the invention of agriculture most people laboured on the land; after the Industrial Revolution they did manual work and moved around under their own steam. But the decline of agricultural labour, plus the invention of cars, labour-saving devices and – most perniciously – home entertainment systems that encourage couch potato behaviour mean we've ground to a collective standstill. This even goes for those who exercise regularly, but spend the rest of the day doing nothing. Strange as it may sound, these people might be less healthy than people who don't exercise but also don't laze around.

Spending too much time inactive is an independent risk factor for all sorts of lifestyle diseases, which means it cannot be cancelled out by exercising at other times. Think of it like smoking: just as you cannot compensate for smoking twenty a day by running ten kilometres at the weekend, a bout of high-intensity exercise does not cancel out the effect of watching TV for hours on end.

The risk appears to come from extended periods of inactivity. Adults who regularly sit for one or two hours at a stretch have a significantly higher risk of early death than those who spend the same overall amount of time sitting but who get up and move every half hour or so.[4]

Long periods of inactivity are bad for our health because they produce a complex cascade of metabolic changes. Unused muscles shrink, and shift from endurance-type muscle fibres, which can burn fat, to fast-twitch fibres that rely more on glucose. Inactive muscles also lose mitochondria, so burn less energy. With the muscles relying more on carbohydrates, unburned lipids accumulate in the blood, which could be why sitting has been linked to heart disease. Fat also gathers in muscles, the liver and the colon – places where it is not supposed to be stored.

And we're paying the price. In the West, lack of cardiorespiratory fitness is now one of the most important risk factors for early death. It accounts for about 16 per cent of all deaths in men and women. Other diseases are also mushrooming. Obesity used to be rare but over a third of US adults are now obese, as are about 17 per cent of US children. And obesity-related diseases are correspondingly on the increase. In 1935, when the world's population was just over two billion, an estimated fifteen million people had type 2 diabetes. By 2010 the population had more than trebled but the number with diabetes had increased fifteen times.

This is a big, big problem. On average, during waking hours, we're sedentary 55 to 75 per cent of the time. Moderate-to-vigorous exercise occupies just 5 per cent or less of people's days.

The answer is to move more regularly. That is all well and good, but for many people it is thwarted by the possession of a desk job. If this applies to you, it is vital to intersperse your day with movement. Even a minute of gentle exercise – a walk to the loo or the kitchen, say – is enough to break up the sedentary periods. As a bare minimum you should aim to briefly leave your desk every thirty minutes. If your job really does keep you glued to your chair, one option is to get a treadmill desk. Failing that, fidget. Do some heel raises, wiggle your toes, roll your shoulders. Every bit counts.

This advice also applies outside of work hours, when the temptation is to flop down and stay put. It is surprisingly easy to rack up several unbroken hours on the sofa, especially if you've been up and at it all day. Train yourself to stand up and walk around the sofa every time there's a natural break in whatever you're watching, reading or doing.

THE TRUTH ABOUT HOW MUCH EXERCISE YOU NEED

So you're sold on the myriad benefits of exercise and the dangers of sloth, and have decided to get fit or up your fitness level. Now what? For many people this is where good intentions start to become a road to hell.

Converting motivation into action is not easy (for more on motivation, willpower and habits, see the next chapter). There's the laziness instinct to overcome. But there's also the problem of knowing what to do, when, and for how long. What even counts as exercise anyway?

If you pay attention to media coverage, exercise can feel like a minefield. As with nutritional advice (is chocolate good or bad for you this week?) it seems to change all the time. One day golf or gardening or sex counts as exercise, the next day it doesn't. On top of that we are bombarded by 'expert' advice and wooed by celebrity workouts that promise miraculous results in no time. The result is to leave people feeling confused, cynical and demoralised.

But if you strip it down, exercise is actually quite simple. In a first approximation, if you're moving around under your own steam you're exercising, and the more of it you do and the more energetically you do it the better.

A useful skill to learn is to regard media coverage with the scepticism it largely deserves. As with nutrition science, most is driven by small-scale studies with unexpected or counter-intuitive results. A dozen studies that find running makes you fitter are not newsworthy. One study that finds that it makes you fatter is (no such study exists, by the way, this is just to make the point). But which is right? Like nutrition science, you need to look at the totality of the evidence.

Similarly, if you're tempted by a celebrity workout regime promising to, say, sculpt a six-pack in six weeks, stop and think. Does it sound too good to be true? If it was that simple, why doesn't everyone have a sculpted stomach? Is it plausible that somebody on TV has discovered a miracle exercise regime that has somehow passed decades of research by sports scientists by?

So how, when and how much should you exercise?

The WHO also advises adults to get at least 150 minutes of moderate physical activity, or seventy-five minutes of vigorous activity (or any equivalent combination, say a hundred minutes of moderate physical activity plus twenty-five vigorous). That is the minimum; ideally you should do more.

This advice does not recognise a distinction between cardio-vascular exercise and resistance exercise, even though these are often seen as distinctive and complementary forms of exercise. But we'll get to that later on in this chapter.

Other guidelines are available. In the US, the Centers for Disease Control and Prevention recommend at least twenty-five minutes of moderate and vigorous exercise a day, or 175 a week.

However, there are good reasons for thinking that even this higher weekly total is woefully inadequate. Because our bodies evolved for hunting and gathering, people who still live as our ancient ancestors did are a good guide to how much exercise we really need.

The Hadza people of the Rift Valley and Serengeti in Tanzania are one such group. They typically get about two hours of moderate to vigorous physical activity a day, mostly in the form of fast walking over hilly terrain, digging tubers from rocky ground, climbing trees, dragging firewood and hauling containers of water. Adults regularly live into their sixties and seventies without any of the age-related diseases we grudgingly accept as inevitable. They have healthy hearts, never develop diabetes, and stay strong and active into old age. These are the rewards of getting the daily dose of exercise that humans evolved to need. In other words, the optimum dose of exercise is two hours of brisk walking a day.

This suggests that current public health guidelines are much too unambitious. Perhaps, then, it is no wonder that around half of adults in the UK don't get enough exercise.

The good news for the exercise-averse is that it's possible to do improving exercise without buying a gym membership or a pair of running shoes. You don't even have to break into a sweat. The definition of moderate exercise includes walking (including around a golf course), gardening, vacuuming and other domestic chores – and vigorous sex.

Exercise intensity is measured in METs, or metabolic equivalents. This is the ratio of your metabolic rate during the activity to your metabolic rate at rest. Moderate activity is defined as anything between 3.5 and 6.5.

So far, so theoretical. Actually measuring your metabolic rate at rest or during exercise requires a trip to a sports science laboratory. But unless for some extremely unlikely reason you need a precise measurement, you can look up the average MET for almost any conceivable activity – from cleaning out gutters to riding a unicycle – on a website called The Compendium of Physical Activities (https://sites.google.com/site/compendium ofphysicalactivities/).

To put 'moderate' into perspective, walking at a regular pace scores 3.5. So do low-intensity household chores such as vacuuming or cleaning. If these hardly seem like exercise, consider that a recent study of more than 130,000 people in seventeen countries found that 150 minutes a week of walking and/or household chores are enough to reduce the risk of early death by 28 per cent.[5]

Vigorous activity – anything above 6.5 – is even better. Slowly climbing the stairs or cycling at a moderate pace lie on the boundary. Running at a gentle five miles per hour scores 8, as does cycling at twenty kilometres per hour; swimming a fast front crawl clocks up 10; and running up the stairs 15.

Weight training (also known as resistance exercise) also counts, though not as much as you might think. A typical set of 8–15 'reps' on a weight machine clocks in at 3.5; a free weights session about 6. However, such resistance training adds to fitness in other important ways, as we shall see later.

People often debate what is the 'best' form of exercise, with swimming and running often touted as candidates. But this is a pointless debate. The 'best' form is what is best for you, and choosing that is a matter of personal preference. It really doesn't matter if you choose Zumba over spin, or decide to run rather than swim. There may be differences in the intensity of the workout and the muscle groups you use but these are negligible compared with the benefits of doing some exercise versus the dangers of doing none. If you're worried about working one set of muscles at the detriment of another, then add in some resistance training on the neglected bits.

Different forms of exercise also carry different risks of acute or chronic injury. If you run a lot, for example, it is worth investing in some advice about how to do it safely. But that is beyond the scope of this book.

Obviously it is better to choose something you enjoy, or at least don't hate. The less complicated and expensive it is, the more likely you'll keep at it. You can run almost anywhere, anytime, but going snowboarding takes planning, money and effort. Joining a club or setting targets can help motivate you. But there will always be days when you have to dig deep.

On those occasions it is worth knowing that even the briefest bouts of exercise are beneficial. Sports scientists used to think that a workout had to last at least ten minutes to do you any good, but are increasingly backtracking. A recent update to the Physical Activity Guidelines for Americans says that any and all physical activity adds up.

There's also the question of how often. Most health authorities recommend doing some exercise every day, or at least spreading it out over the week. But about half of people who exercise regularly cram all their weekly exercise into one day. Data from more than 63,000 adults in the UK shows that people who opted for this 'weekend warrior' regime had pretty much the same reduced risk of early death from all causes, including cardiovascular disease and cancer, as those who exercised little and often.[6] Even weekend warriors who did less than the recommended amount for the week fared better than inactive people. So if you have let the days go by without doing any exercise, don't write the week off.

And even if you've fulfilled your target, that's no reason to stop. It is possible to over-train but that is a risk for elite athletes, not ordinary people exercising recreationally. You should also give yourself time to recover properly – let fatigue and stiffness be your guide – and be wary of injuring yourself. But these caveats notwithstanding, there is no such thing as too much exercise.

Can exercise kill me?

Like it or not, during vigorous exercise the risk of a heart attack does increase. Couch potatoes are fond of citing the case of Jim Fixx, who is credited with sparking the running craze in the US in the 1970s. In 1984, at the age of 52, he dropped dead from a heart attack midway through his daily run. The size of the increased heart-attack risk depends heavily on how accustomed you are to that exercise. For someone who is completely unfit and goes for a run, the risk can be a hundred times higher than when at rest. This risk lasts for the duration of the exercise and up to half an hour afterwards. Even for someone who runs five times a week, the risk roughly doubles. But given that the risk of a heart attack at any given moment is very low, even these elevated risks remain small.

And in any case the raised risk pales into insignificance compared with the overall lifetime benefits of regular exercise. Study after study has shown that staying active lowers the risk of a heart attack by 50 to 80 per cent. The sensible advice is to be sensible. If you are unfit but want to get fit, start gradually and build up. It may be worth getting a medical check-up too; Fixx's autopsy revealed that he had advanced atherosclerosis, probably hereditary.

What if exercise just isn't for me?

While physical activity is necessary to get fit, it doesn't work for everybody. There is variation in how people respond to exercise, and some people hardly respond at all.

The classic research in this field is the Heritage family study, begun in the 1990s. Researchers recruited 481 sedentary people from ninety-eight families, and subjected them to a twenty-week training programme while monitoring their progress – or lack of it.[7] While most participants' aerobic fitness improved dramatically, others showed a smaller response and about one in ten showed no change whatsoever, despite doing forty-five minutes of vigorous exercise three times a week.

It turned out that being a 'low responder' runs in families and is probably down to genetics; ditto being a high responder. If you think you are one of those people who cannot get aerobically fit regardless of how much you sweat, don't give up.

Even those whose aerobic fitness did not increase saw improvements in blood pressure, cholesterol, insulin response and abdominal fat. There's no such thing as a complete non-responder.

THE TRUTH ABOUT 10,000 STEPS

In the age of personal fitness-tracking technology, exercise targets are more often couched in terms of the number of steps each day, with 10,000 usual. In the modern world that is quite a challenge. People in the US and UK, on average, take about 1,500. But the research on the Hadza suggests it is not nearly enough.

Two hours of brisk walking a day equates to about 15,000 steps. Evidence that this translocates to Western lifestyles comes from a study of postal workers in Glasgow.[8] Those who racked up more than 15,000 steps a day carrying the post have cardio-metabolic health on a par with the Hadza.

This also tells us that the 10,000-step target – originally a marketing ploy dreamed up by a Japanese manufacturer of pedometers in the 1960s – is less than the optimum amount. In fact, it probably gives an amount of exercise comparable to the recommended twenty-five minutes. Hang on, you might think, if 15,000 steps is the same as two hours' vigorous walking, then 10,000 should be about eighty minutes. But many of the steps we take in our modern lives are low-intensity, and count for less. The Hadza traverse hilly, rough terrain searching for food. We amble along flat city streets and down supermarket aisles. On a fitness tracker 10,000 steps roughly equate to 3,000 hunter-gatherer steps. For reference, climbing four flights of stairs clocks up fifty hunter-gatherer steps, a one-kilometre brisk walk 1,250 and half an hour of running 3,500.

Given that, 10,000, let alone 15,000, moderate-to-vigorous

steps a day may seem a distant goal, but don't be discouraged. One step is better than none. Studies consistently show that even modest amounts of exercise produce significant health benefits compared with zero.[9]

For an average person who is not an elite athlete, and hence not in danger of running themselves into the ground, the only dangerous dose of exercise is zero. Your goal should be to do as much exercise as you realistically can in a way you enjoy. The best exercise is the one that keeps you coming back for more.

THE TRUTH ABOUT FAST FITNESS

One of the main reasons people give for not exercising enough is lack of time. That is often just an excuse, but if you really are pressed for time an option to consider is HIIT, which stands for high-intensity interval training. It promises to get you fit in as little as ninety seconds, which sounds too good to be true. But sports scientists say it works – as long as you stick to the regime.

Researchers at the National Institute of Fitness and Sports in Japan invented HIIT in the 1990s. They showed that a six-week course of four-minute workouts, comprising repeated cycles of twenty seconds of flat-out exercise followed by ten seconds of lower-intensity work, done four days a week, brought greater improvements in aerobic fitness than an hour's normal workout done five days a week.[10]

HIIT has since been adapted into a wide variety of regimes. Some last just ninety seconds, others a gruelling half an hour or until complete exhaustion, whichever comes first. The intense exercise is often cycling on a fixed bike but can be anything vigorous: running, rowing, swimming, climbing stairs. The goal is to get to at least 80 per cent of your maximum heart rate.

The rest period is often a gentler version of the exercise, or total rest. What most of the regimes share is that each repetition has a two to one ratio of intense and moderate exercise.

A large body of evidence shows that HIIT is as good as, or sometimes better than, moderate-intensity training for cardiovascular and metabolic health. It is especially good for people who are overweight or obese, quickly improving their blood pressure, lung function and fasting glucose levels, and in the long term (done for more than twelve weeks) their waist circumference, levels of body fat and resting heart rate.

But beware: done properly, HIIT is hard graft. Eight twenty-second bouts of maximum intensity exercise with ten seconds rest in between sounds manageable but is enough to make many people physically sick afterwards.

But you don't have to torture yourself. Longer rest periods don't seem to cancel out the benefits. One regime involves doing a total of sixty seconds of all-out sprinting within a ten-minute window of jogging, and has been found to be as effective as jogging continuously for fifty minutes. Simply incorporating some bouts of intense exercise in a longer routine – sprints or steep hills while running or cycling, say – will deliver some of the benefits. This is where HIIT blends into regular interval training, which is a proven method that has been used for decades by athletes to improve performance.

THE TRUTH ABOUT
CARDIO AND RESISTANCE

Exercise regimes usually make a distinction between two kinds of exercise – cardio and resistance (aka cardiovascular and weight training) – and stress the importance of doing both. Cardio is exercise that increases your heart rate while resistance is basically muscle strengthening.

This dichotomy is not entirely justified; there is no such thing as 'pure' cardio or resistance exercise. Any form of cardio-vascular exercise works muscles at the same time, and muscle-strengthening exercises increase your heart rate (recall that weight training counts as moderately vigorous exercise). But nor is it entirely false, and the latest evidence suggests that doing both kinds of exercise is more beneficial than just doing one.

Several studies have suggested a link between muscle strength and living longer, but for a long time it was unclear whether muscle strength per se was the important factor. People who are muscular are also likely to be aerobically fit, lean and generally healthy – all features known to extend lifespan.

However, some large, well-designed studies have settled the question. Even when factoring out aerobic fitness, muscle strength is correlated with lower mortality.[11] The bottom line is that both aerobic fitness and strength make independent contributions to health. In recognition of that the American College of Sports Medicine recommends two sessions of strength training a week, consisting of about ten repetitions of strengthening exercises for all ten major muscle groups.

Should you work your core?

Yes. Probably. Your core is loosely defined as the muscle groups in your torso, which are involved in almost every conceivable bodily movement. Many exercise regimes – especially Pilates – explicitly work on the core muscles, often painfully so, and gym instructor types often claim that having a strong and stable core improves posture, resistance to back injury and performance at pretty much any kind of physical activity. However, the scientific literature is very mixed, with no consensus on the true benefits of core strength and stability, how best to achieve it, nor its knock-on effects on athletic performance. But having stronger and fitter muscles is generally a good idea so, unfortunately, a bit of core training should be part of a balanced exercise regime. It will hurt, but if it ain't hurting it ain't working. So get planking.

What about yoga and Pilates?

Based on the standard measure, METs, most forms of yoga and Pilates are not aerobically intense enough to qualify as moderate forms of exercise, though power yoga sneaks over the line with a count of four. So if you want to get aerobically fit there are better options. But yoga and Pilates have other benefits, including building strength and suppleness.

THE TRUTH ABOUT WORKING UP A SWEAT

Exercise is a bit like genius: 1 per cent inspiration and 99 per cent perspiration. If you're going to get fit, it makes sense that you should aim to work up a really good lather. Finishing a workout drenched in sweat certainly makes it feel like you've achieved something.

But feelings can be deceptive. The amount you sweat is not a good indicator of the effectiveness of exercise, but simply how much heat your body needed to dissipate. But it can be a rough guide to your fitness level. Perhaps surprisingly, the fitter you get, the more you sweat. You can forget the idea that sweating more leads to weight loss – you may be a bit lighter straight after a session, but it's caused just by dehydration.

Aiming for maximum perspiration can also be counterproductive. Sweating causes water to be lost from blood plasma, which causes a drop in blood volume and means your heart has to work harder to deliver oxygen to muscles. That can be a problem. An average person sweats between 0.8 and 1.4 litres per hour during exercise. The International Olympic Committee says that performance is likely to be impaired by dehydration if you lose 2 per cent of your body weight or more through sweating – a litre and a half for someone weighing seventy-five kilograms. So sweating heavily for more than an hour may reduce the effectiveness of your workout.

At some point you will need to replace the lost fluid, but unless you've sweated heavily for an hour or more and intend to carry on you can safely do it after your workout. The idea that you will get dangerously dehydrated during exercise and need to 'push the fluids' owes more to sports drinks marketing than science. As at other times, let thirst be your guide. And don't be fooled by the hype about super-hydrating drinks. Water will do the job perfectly well.

One big claim for sports drinks is that they are full of vital minerals like sodium that are lost in sweat. This is undoubtedly true, but that does not mean you need expensive and often sugar-laden sports drinks to replenish them. You'd have to sweat an awful lot to lose physiologically relevant amounts of sodium. Most diets include more than enough sodium for the casual exerciser, but if you are going for extremely long periods, such as an ultra-marathon, a drink that tops it up could be useful.

THE TRUTH ABOUT
STRETCHING AND INJURY

Accepted wisdom has it that any workout should start with a warm-up – a period of low-intensity exercise and stretching to prepare your body for the rigours ahead – and end with a warm-down of mostly stretching. The rationale is that the exercise part warms and loosens up your muscles and joints and gets your circulation going. Stretching, meanwhile, guards against injury during vigorous exercise, and soreness and stiffness afterwards.

There's no doubt that warm-ups do improve performance, but stretching probably doesn't achieve as much as you think. The evidence shows that regardless of whether you do it before or after exercise, stretching does little except fractionally reduce the risk of injuries to muscles, ligaments and tendons.

Injuries notwithstanding, it almost certainly won't help with muscle soreness. This comes in two forms: acute, which is basically muscle fatigue during and shortly after exercise, and delayed onset, or DOMS, which is a sign that you have done some minor damage to the muscle tissue. Acute muscle soreness should ebb away in a few hours. DOMS, in contrast, kicks in twenty-four to seventy-two hours after exercise and can last up to a week. It can be extremely debilitating.

If you ask somebody about the cause of muscle pain due to exercise, they will probably say something about the build-up of lactic acid. It is true that lactic acid is produced during intense exercise as a by-product of anaerobic respiration – which happens when oxygen is in short supply but energy is still required. Lactic acid was once thought to be a cause of muscle fatigue, but that is now controversial.

In fact, there is growing evidence that lactic acid can also be an extra source of fuel. Muscle cells can reuse lactate by transporting it from the cytoplasm – the fluid that fills cells – into the mitochondria. Endurance training seems to increase the amount of lactate that is taken up by mitochondria.

Lactic acid is definitely not the cause of DOMS, which happens because of minor, reversible damage to muscle fibres. These are designed to act like a ratchet, 'walking' past one another as the muscle contracts and sliding back again as it relaxes. But as they become energy-depleted through continued exertion, the ratchet mechanism fails and the fibres can be ripped apart, causing micro-tears in the muscle.

Once this damage occurs, recovery and repair processes kick in. This includes inflammation, which is probably the main cause of the pain and stiffness. There's not a lot you can do about it except take painkillers and wait for it to go away. Massage can help a bit but ice, stretching and homeopathy don't do anything. Further exercise can reduce the pain but this is a bit like hair of the dog for a hangover – you'll pay double later.

DOMS is considered a sub-clinical injury that does not need medical treatment and will resolve itself. But severe pain is a warning sign that you should dial it back a bit, or risk a proper muscle injury.

However, DOMS is worth enduring because as the muscles are repaired, they grow stronger and adapt to the greater burden, increasing the size of the fibres and building up strength in the muscle. The next time you do the exercise that left you in pain, you will find it marginally easier at the time and marginally less painful later. This is how exercise causes muscles to grow. No pain, no gain.

But forget the idea of toning. Muscles get bigger and smaller but they don't tone up or down. The toned look comes from having bigger muscles obscured by less subcutaneous fat. Both can result from exercise but there is no specific exercise for toning. A related and equally bogus concept is spot reduction, which claims that exercising a localised muscle group such as the abdominals will burn off fat in that area. It won't.

Another myth is that having more muscle mass will significantly increase your basal metabolic rate, meaning you are burning calories like billy-o even when resting. It is true that a resting muscle consumes energy, but much less than many fitness gurus will tell you – about ten to fifteen kcal/kg per day. A resistance exercise programme can add two kilograms of muscle, which will increase basal energy expenditure by no more than thirty calories a day – less than in a single cream cracker.

Fat tissue also consumes energy, albeit at a slower rate than muscle. So if you lose fat at the same time as gaining muscle the difference in your resting metabolism will be negligible. It is true that, per unit volume, muscle does weigh more than fat but this is unlikely to tip your BMI into an unhealthy category unless you are very muscly.

You can learn to love or at least appreciate muscle soreness, but exercise also carries the risk of an injury, mostly pulled muscles and twisted joints but occasionally something much worse. Around half of regular runners and players of team sports such as football get some kind of musculoskeletal injury every year.

It's hard to know whether to rest an injury or push on through the pain. As a rule, if you are taking up a new activity, you'll endure some aches and pains at first. Parts of your body that are not used to exercise will complain, but adapt. But if the pain becomes debilitating, give it a break and see a doctor. And if you decide to resume once it has healed, get some advice on technique.

THE TRUTH ABOUT MEASURING FITNESS

'Fitness' is technically defined as aerobic fitness, or how effective the body is at delivering oxygen to muscle cells.

The best way of assessing it is to measure your VO2max – the maximum rate at which your body can transport oxygen to your muscles and then use it to convert fuel into energy. It is measured by the volume of oxygen you consume per kilogram of body weight per minute. To measure it, you run on an accelerating treadmill while breathing through a mask to gauge the oxygen level of the air you breathe out. The higher your VO2max, the fitter you are.

The average healthy male has a VO2max of between forty and fifty millilitres per kilogram per minute, which may climb to around sixty to sixty-five after prolonged training. Elite endurance athletes like Tour de France cyclists have a VO2max over eighty. Bjorn Daehlie, a Norwegian cross-country skier who holds the record for the most medals won at the Winter Olympics, reportedly had a VO2max of ninety-six – the highest ever recorded.

There are also ways to estimate VO2max that don't require a sports science laboratory. The only equipment needed for the Rockport Fitness Walking Test, for example, is a weighing scale and a watch. Time how long it takes you to walk a mile (1,609 metres) on a flat surface – preferably a running track for accuracy – as quickly as possible, then measure your heart rate. Plug the time and heart rate, along with your age, gender and weight, into the appropriate equation or find a website that will do it for you – try the Brian Mac Sports Coach site at www.brianmac.co.uk/rockport.htm – and you'll get a ball-park value for your VO2max.

What counts as a good score varies quite a lot by age and gender, but for an adult male anything above forty is better than average and for an adult female anything above thirty.

THE TRUTH ABOUT
EXERCISE ADDICTION

One of the joys of exercise, and a reason why some people get so hooked on it, is the runners' high. This is a state of euphoria and intense mental clarity induced by a long and gruelling bout of exercise (not just running), caused by the release of natural drug-like molecules in the brain. These include the opioid-like endorphins and the endogenous cannabinoids, which are similar to the active compounds in marijuana.

Exactly how long and hard you need to exercise to achieve the high varies enormously from person to person and workout to workout. The high is probably an evolved response from a time when endurance running was vital to our survival. Humans are physically adapted for long-distance running, which has led evolutionary biologists to speculate that we once made a living as endurance hunters, chasing prey until it collapsed with exhaustion. The high evolved as a reward to keep us running even when knackered, to keep us focused and to mask pain.

There are downsides to the high, however. Like the drugs it mimics, it can become physically addictive, potentially leading to injuries and an illness called overtraining syndrome. Sports scientists recognise exercise addiction as a genuine addiction, albeit a rare one. People can also become addicted to exercise because of an unhealthy fixation on body image.

As with other addictions, defining and diagnosing it is difficult. But in general, if exercise becomes the sole focus or pleasure of your life and starts getting in the way of social, family and work commitments, seek help.

THE TRUTH ABOUT EXERCISE AND WEIGHT LOSS

The biggest motivating factor for many people to exercise is to lose weight, or avoid gaining it. And there are good reasons besides vanity for wanting to do so. Being overweight or obese is associated with a range of life-limiting conditions including cardiovascular disease, cancer and type 2 diabetes.

There's no great mystery about how to achieve weight loss. It is ruled by the simple and unbending rule of CICO – calories in must be less than calories out – which we met in the section on calorie counting (see page 113). It is why both dietary restriction and exercise are key to staying trim. The first cuts CI, the second increases CO.

But exercise alone is a surprisingly ineffective way of shedding weight. Numerous studies have shown that people who exercise to lose weight rarely achieve it without dieting too.[12]

This may sound familiar to anyone who has tried to pursue an exercise-led weight-loss regime. Despite tipping the scales in favour of calories out, the actual scales don't budge, or keep on creeping in the wrong direction. This has come to be known as the exercise paradox. Scientific understanding of it is still in its infancy, but what we do know can help you to achieve your weight-loss goals.

Until recently, the paradox was explained away by the logic that exercise makes you feel more hungry (and more virtuous) so you overcompensate by eating too much. It is tempting to think that, having done a workout, you have licence to eat and drink whatever you like. You can't. As the old adage says, you can't outrun a bad diet.

It doesn't help that people hugely overestimate the amount of energy they use during exercise and similarly underestimate the calorific value of food. Consider that running a mile will burn roughly 120 calories – about the same as in a single slice of brown bread or a slice of cheddar. If you run four miles you've earned two cheese sandwiches.

In one study, people who did a treadmill run were asked how many calories they thought they had burned. They estimated 800, even though they had only gone through 200.[13] Offered a buffet to regain the energy they had run off, they ate on average 550 calories. There's also the fact that you can consume so fast what it takes hours to burn. The run may take you ten minutes – the cheese sandwich is gone in seconds.

This is backed up by longer-term studies. Over the course of a year, 150 minutes per week of moderate exercise burns roughly 50,000 calories. That's about the same as in six kilograms of fat. Yet people who do it weigh only two kilograms less on average than the sedentary.

Another experiment found that exercise can actually be associated with weight gain.[14] It kitted out nearly 2,000 people with activity monitors to gauge their habitual pattern of activity. Their weight was then tracked over several years. On average people who did the recommended 150 minutes of moderate exercise per week gained more weight than those who did less.

This experiment also revealed that compensatory eating cannot be the whole story. Analysing the data in more detail revealed, surprisingly, that people who exercised over and above the recommended levels did not burn significantly more calories than people who did the bare minimum.[15] Even more surprisingly, heavy exercisers only burned about 200 calories a day more than totally sedentary people. It seems that once you go beyond the 150 minutes per week mark, calorie expenditure plateaus. That cannot be explained by compensatory eating, as the researchers were looking specifically at the calories-out side of the equation. So how can it be that people who do more exercise don't expend more energy?

Some clues come from research on those lean and healthy hunter-gatherers, the Hadza people of Tanzania. A few years ago, scientists studied their metabolism. They weren't looking for weight-loss wisdom but just wanted to know how many calories it took to be a hunter-gatherer. They expected the answer to be 'a lot', but they were surprised. The Hadza burned through only slightly more energy than Westerners who spend most of the day sitting, with men using up about 2,600 calories and women 1,900.

The findings were hard to reconcile with the laws of thermo-dynamics. Moving about uses up energy, so the more you move, the more you should burn. What was going on? It turns out that the laws of thermodynamics are safe. The explanation is that after a period of hard physical graft, we unconsciously compensate by grinding to a standstill for the rest of the day. This is seen in mice with access to a running wheel. The more they use the wheel, the less they move around afterwards, saving almost exactly the same number of calories as they burned on the wheel.

People make similar adjustments, though they are unaware of it. After a hard morning workout, for example, people reduce energy expenditure in the afternoon, resulting in a roughly equal total calorie expenditure on days with and without exercise. Another study found a dialling down of general activity for six days after a hard workout.[16]

One of the key contributors to background energy expenditure is fidgeting, and this is one of the activities we cut back on after a workout. It is therefore a mistake to count calories burned during a workout but disregard those used during the rest of the day. Normal activity (fidgeting included) actually makes up the lion's share of your daily energy expenditure and unconsciously cutting back on it can cancel out any gains made from exercise.

Fidgeting can burn a surprisingly large amount of energy. In one experiment, researchers got volunteers to eat an extra 1,000 calories per day for eight weeks while their weight and movement were monitored.[17] For those who kept the weight off, the key turned out to be fidgeting and other subtle movements, even as small as posture control. Other research backs this up – all the jiggles and wriggles of the most fidgety among us burn around 800 calories a day.

You might think that the answer is to cultivate your inner fidget and avoid idling, but unfortunately your metabolism is playing the same game. When you exercise hard, your body adapts and slows down your metabolic rate. This is called 'metabolic compensation': the more energy you burn from exercise, the less energy your body uses for basic functioning. Doing more exercise at this point is not going to eat into your energy budget. This is why athletes who over-train can succumb to over-training syndrome, a constellation of problems including reduced immunity and fertility. White blood cell counts crash, infections are hard to shake and women stop ovulating. Exercise stops being healthful and starts being harmful. This might explain why extreme exercisers have slightly higher mortality rates than people who work out a few times a week.

Ultimately, it is hard to avoid the conclusion that diet offers greater potential than exercise to tilt the CICO equation towards weight loss.

But exercise still has its place. As well as making you fitter, if you do lose weight, it can help prevent the common problem of putting it all back on. In studies of successful dieters, ongoing weight loss is poorly correlated with the amount of exercise. However, in the long run, those who increase physical activity the most regain the least weight or keep it off.

Burning fat

One supposedly desirable destination hidden somewhere in the gym is the 'fat-burning zone' – a level of exercise precisely calibrated to get rid of your flab. Even more temptingly, this zone is supposedly arrived at not by working your butt off but by taking it steady. The source of this idea is that the body can quickly turn carbohydrates into fuel, so they are called upon in an intense workout, whereas fat is burned more slowly, meaning that it is called upon during less intensive exercise.

Unfortunately, it doesn't add up. At very low exercise intensities we do burn proportionally more fat than carbohydrate, but you'd have to work out very gently for a very long time to burn a significant amount. If burning fat is your goal, you're much better off working out at higher intensity. You'll burn through carbs, for sure, but you'll also use fat at a higher rate than you can achieve at walking pace.

THE TRUTH ABOUT MOTIVATION

Oscar Wilde hit the nail on the head when he wrote: 'I can resist everything except temptation'. For most of us, the main obstacle standing between our current selves – probably a bit unfit and slightly flabby – and the athletic and slender person we'd like to be is a lack of willpower. Our trainers and gym kit sit in the wardrobe while we idle in front of the TV. We tell ourselves that we will be stronger tomorrow. Tomorrow, we cave in again.

Psychologists used to have a good explanation for why humans are so feeble in the face of temptation. Willpower is a limited resource that you only have so much of each day. Once you have used it up – perhaps by willing yourself to do a tedious task at work, resisting the temptation to eat chocolate or gritting your teeth and going to the gym – there's nothing left. At that point you are a sitting duck.

That is a rather demoralising conclusion. We are so surrounded by unhealthy temptations that our resources will surely run dry before the day is done. But recently psychologists have revised their understanding. Willpower is not finite, but a renewable resource that can be replenished as you go along. In fact, your willpower is only limited if you think it is. This means that if you cultivate the right mindset, extraordinary levels of motivation and self-control are yours for the taking.

The ability to delay gratification in pursuit of longer-term goals is known to have a big impact on people's lives. In 1972, Walter Mischel devised the 'marshmallow test', which has become a classic in psychology. To test the willpower of children aged four to six, he put a treat – a marshmallow, an Oreo cookie or a pretzel – in front of each child and left the room, telling them that they could eat it if they wanted but if it was still there when he came back fifteen minutes later, they would get two. A minority of children ate the snack as soon as he left the room but the rest at least tried to resist. About a third of them succeeded. These children went on to be more successful in life, both in education and work.

More recently, psychologists who followed 1,000 people in New Zealand for decades discovered that those with the least self-control as young children were more likely to experience unemployment, poor health and criminality as adults.[18]

The orthodox view of willpower as a finite resource – known as ego-depletion theory – came from experiments done in the late 1990s. These repeatedly showed that people who have exerted willpower in some way perform relatively poorly on subsequent tests of self-control. For example, give them a difficult puzzle that requires persistence and they will do badly on a second, different type of tough puzzle. People given an easy initial puzzle don't show this decline in performance.

In another classic experiment, volunteers were shown into a room with a plate of freshly baked cookies and a bowl of radishes on the table. Some were told they could eat whichever they wanted, others that they could only eat the radishes. The subjects were then left alone for a few minutes, before being given an unsolvable puzzle. Those who had been denied the cookies, and hence had to use willpower to resist temptation, gave up more easily.[19]

More than a hundred other studies have shown this same pattern – tax somebody's mental energy in some way and their ability to resist temptation diminishes. The researchers concluded that willpower is like a muscle that gets fatigued with overuse.

Psychologists went on to suggest that the cause of willpower fatigue is depletion of glucose through hard mental graft. Giving participants a sugary drink between challenging tasks can restore willpower. In 2012, the American Psychological Association published a guide on the psychology of willpower that recommended eating regular snacks.

This may partly explain why dieting is so hard. To resist food requires willpower, but to have willpower we need to eat. Ironically, the very goal of dieting robs us of the strength we need to succeed.

It turns out we spend a lot of time depleting our willpower reserves. Another study kitted out 200 adults with beepers and asked them to record what they were doing, including whether they were currently resisting temptation, when their beeper went off. This revealed that people typically spend an incredible three to four hours a day exerting self-control.

Perhaps, then, it is no wonder that we cave in so frequently. But more recent research presents a more nuanced picture that suggests we are not slaves to ego depletion.

The new thinking is related to mindset, which refers to how people think about their own abilities and how they respond. People with a 'fixed' mindset tend to see abilities such as intelligence as set in stone, and give up easily. But people with a 'growth' mindset see them as malleable, and are more resilient and persistent. These mindsets can be taught and are sometimes used in education to boost performance.

When psychologists combined the willpower experiments with mindset tests, they found something very encouraging. Before doing the tasks the volunteers were asked whether they considered willpower to be a limited resource, depleted by effort, or potentially unlimited. Those who said 'limited' showed the usual ego-depletion effect, whereas the others were much more persistent.[20]

Further studies have discovered that willpower can be boosted just by telling people that it can.[21] When people were shown statements such as 'it is energising to be fully absorbed with a demanding mental task', they continued to get better through a difficult twenty-minute memory challenge. Another group told willpower is limited flagged about halfway through.

As yet the mindset element has not been applied to tests of temptation to eat food. But from what we know, a simple change of mindset could allow you to tap into almost unlimited reserves of willpower and self-control. Tell yourself that resistance is not futile, and you can do today what you usually put off till endless tomorrows.

One way to supplement your willpower is to try to form good habits around health behaviours. Habits are behaviours that we perform automatically to save mental resources. If you had to concentrate on every routine task, such as getting dressed or cleaning your teeth, life would be nearly impossible. Habit formation is such an efficient use of resources that as much as 40 per cent of our daily behaviour is habitual.

To understand habit formation, think about learning a tricky physical task such as driving a car or playing a musical instrument. At first you have consciously to think about every movement, but gradually they are delegated to your subconscious until you can perform them automatically.

Habits – both bad and good – form in a similar way. They start as what psychologists call 'goal directed' conscious behaviour such as popping into the corner shop on the way home from work and buying beer. But do it repeatedly and soon enough the action becomes habituated and you're buying beer most days whether you consciously want some or not.

Annoyingly, bad habits seem easier to form than good ones. But that may just be because we are more aware of our bad ones. Chances are that you habitually brush your teeth morning and night, get dressed before you go out of the house, let passengers off the train first, avoid farting in public and obey all sorts of other social conventions without agonising about them.

As far as the brain is concerned there is no distinction between good and bad habits. At times of stress people tend to fall back on habitual behaviour. During exam times, for example, students often find that their bad habits, such as unhealthy snacking or nail biting, increase.[22] But so do good habits, like exercising.

This is where willpower and habits intersect. When willpower is weak, habitual behaviour takes over. So to maintain a healthy lifestyle it is a good idea to cultivate good habits and eliminate bad ones.

Easier said than done, but there are some pointers from science. Recall that habits start out as goal-directed conscious behaviour. If you want to cultivate a habit of, say, going running at the weekend, you're going to have to start by consciously putting on your trainers and going for a run at the weekend. But over time this will become more and more habitual until you find that you just do it.

Ditto healthy eating. If you want to eat more fruit and vegetables, you will have to start by deliberately eating more fruit and vegetables. But if you eat a piece of fruit after every meal, soon enough you'll do it without thinking.

One useful technique is to try piggybacking it on something you already do, for instance, dropping into the gym on your way home from work.

Habits are also highly contextual; the sub-routine can be initiated by environmental cues or places. In one experiment, researchers showed people a movie in either a cinema or a conference room. In both settings, the participants were given either fresh popcorn or week-old popcorn. In the cinema people ate more of both, even though they said they didn't like the stale popcorn. The context cued a snacking habit they had built up from years of eating popcorn in cinemas.

You can work this to your advantage. On Saturday night leave your running shoes where you'll see them on Sunday morning. The mere sight of them will help to propel you out of the door.

It's rarely enough to form good habits, however – it's just as important to break bad ones. The trick is to disrupt the subconscious routine. Make an effort to not go past the beer shop on your way home. Take the chocolate and crisps from their usual place and put them somewhere else, so when you mindlessly open the cupboard they are not there. Yes, you could easily go and get them but at least you'll have a conscious moment to decide not to.

The contextual nature of habits can also help to break bad ones. The optimum time to try is when your routine changes, such as when starting a new job, moving house or going on holiday.

Another tip is not to worry about lapses. A study that followed a hundred people as they tried to form new habits found no long-term consequences from slipping up for a day here or there. So if you exercise daily for a month, but then miss a couple of sessions, don't take that to mean you've failed.

Time-wise, folk wisdom says that it takes twenty-one days to make or break a habit. The science suggests this is a bit optimistic and the average is around three months. But there is wide variation. Studies of habit formation suggest it can take from eighteen to 254 days. So don't give up; you'll get in the habit in the end.

THE TRUTH ABOUT
A POSITIVE MINDSET

If exercising really does feel too much like hard work, there's another weapon you can deploy: your mind. Believe it or not you can think yourself fitter, slimmer and more youthful. All it takes is to cultivate a positive mindset.

That might sound like New Age mumbo-jumbo but is backed by sound science. It all comes down to something called the placebo effect.

Placebos are sham treatments used in medical trials to test how effective (or not) the real treatment is. In a drug trial, for example, half of the participants are given the actual drug and half an identical-looking sugar pill. Nobody – not even the doctors doing the trial – knows who gets what.

The placebo should do nothing but often causes measurable effects, triggered simply by people's expectations. The effects can be positive, such as an easing of symptoms, but can also be negative, producing side effects such as nausea and rashes. This is the nocebo effect, the placebo effect's 'evil twin'. The effects can occur even when people know they are taking a placebo.

The placebo effect is often exploited in medicine, but is now being explored as a way to improve general health and well-being. And it is something you can use on yourself.

One influential experiment involved eighty-four hotel cleaners, who do a lot of physical exercise at work but often don't realise it. This, researchers hypothesised, might be preventing them from seeing the full benefits.[23] To manipulate their mind-sets, half of them were given detailed information about the physical demands of their work and informed that they were meeting official US exercise recommendations. A month later these cleaners had lost weight and their blood pressure had fallen. The other cleaners showed no difference.

Another study examined data from existing health surveys of more than 60,000 people.[24] It found that the participants' 'perceived fitness' was a better predictor of their health than the amount of time they actually spent exercising. Overall, people who had a more negative view of their own fitness were about 70 per cent more likely to die during the survey, whatever their exercise routine.

It is possible to take advantage of this mind over matter. Don't deceive yourself about your fitness and health, but make sure you feel positive about the healthy behaviours you do manage.

Mindset can also help to improve your relationship with food. In another experiment, volunteers drank a milkshake, and then had levels of their 'hunger hormone' ghrelin measured.[25] Ghrelin normally drops after a meal. Although everyone drank the same milkshake, some were told it was low calorie and others that it was full fat. Those who thought they had drunk a diet shake had higher levels of ghrelin afterwards and felt less full.

Similarly, volunteers given a sugary sports drink responded to it more healthily if they were told it was a new type of drink that solidifies in the stomach.[26] Normally sugary drinks don't trigger a fullness response and just add 'empty calories' to the diet. But the people who thought the drink had solidified (it hadn't) responded as if they had eaten solid food. They had lower ghrelin levels, a healthier insulin response to the sugar, felt fuller and ate less later on. This is something you can also act on. When dieting, cultivate what researchers call a 'mindset of indulgence', savouring whatever you are eating as if it were a real treat.

Perhaps the most useful research concerns ageing, with

evidence that a positive mindset can add years to your life. In a famous experiment done in 1981, researchers took a group of pensioners to a monastery in New Hampshire and told them to act as if they were twenty years younger. The place was decorated as if were the late 1950s and filled with music, magazines and books from that era. There were no mirrors, only pictures of the pensioners when they were younger. After five days, their arthritis had eased, their postures were more upright and their performance on IQ tests better.

Inspired by this study, other teams have shown that attitudes can influence how our bodies age. Overall, people who view ageing positively even before it affects them live seven and a half years longer than those who associate it with frailty and senility.

A positive mindset is not a panacea. But these findings could help all of us benefit more from efforts to achieve a healthy lifestyle through exercise, diet and sleep.

As a teenager I was extremely active, playing football almost every day and going cycling with my mates at the weekend. But a serious illness – and the subsequent discovery of beer, cigarettes and nightlife – meant that by my mid-twenties I was almost totally inactive. I remember being slightly scared of exercise, kidding myself that I was still fit but knowing deep down that I wasn't.

One day in my early thirties I pulled some old trainers out of the wardrobe and went for a run. It was a chastening experience – I probably managed half a mile before collapsing in exhaustion – but I felt good and virtuous afterwards. I did it again, and then again, and I quickly upped my distances and regained some fitness. I soon found I was running twice or three times a week and feeling antsy if I didn't run. I bought a bike and started cycling too. Nowadays I cannot imagine going back. I ran over 400 miles last year, cycle to work most days, and (mostly) love it. I don't know how fit I am, but I know I'm much fitter at fifty than I was at twenty-five.

This is all anecdotal, of course. But the science backs me up. If there's just one message that you take away from this book, it is this: GET MOVING. Exercise is a cure for so many ills that if available in pill form, doctors would feel compelled to prescribe it to all of us. Any amount of exercise is better than none, and almost any activity that involves moving around under your own steam counts. I'm living proof that, even if you have got out of the habit of exercising, getting back in the saddle isn't as hard as you think it is going to be and the rewards are rapid and self-reinforcing. As a manufacturer of sports equipment is fond of saying, just do it.

THE TRUTH
ABOUT STAYING
WELL

PREVENTION IS BETTER than cure, or so we're told. Medical authorities certainly seem to agree. Preventive medicine is more popular than ever, and 'wellness' (by which is meant absence of illness) is becoming a key goal of public health policy.

If you have been to see a doctor recently, there's a good chance that whatever your specific complaint, you also got a general check-up: weight, blood pressure, cholesterol and perhaps some other tests. For many people that ends with a prescription for a condition they didn't know they had or were at risk of developing – perhaps a statin to lower cholesterol, or an ACE inhibitor for high blood pressure. Often, they will be taking those pills for the rest of their lives.

You may also have been offered some kind of screening, for breast or bowel cancer, say, in a bid to catch the disease early and hence increase your chances of beating it. Such early interventions or prophylactic measures seem like a good thing. Where's the harm in catching potential problems early and using modern medicine to nip them in the bud, or in dishing out pills to prevent them in the first place?

But care is needed. The anti-cholesterol drugs called statins certainly work but they can cause side effects, and we don't know much about the risk of taking them long-term. Routine screening for diseases, including some cancers, once seemed a sure-fire route to saving lives. But overall, most mass-screening programmes proved to be ineffectual or even harmful and were duly dropped. Over-screening is a real problem: false positives lead to unnecessary medical intervention and psychological trauma, while false negatives can lead people to ignore genuine symptoms. And then there's the issue of over-medication, where mild conditions such as slightly elevated blood pressure are deemed to require drugs.

This is the topsy-turvy world of preventive medicine, where measures that look like common sense often turn out to be counterproductive, and others that look pointless are not. This chapter will take a good look at which ones are worth it, and which should be avoided until there's actually something to be cured.

We'll also see how to avoid some of the more insidious health hazards that you've probably not given enough thought to: stress, air pollution, lack of sunshine, nature deprivation and loneliness. Your doctor may not give you advice about them, but if you're interested in a preventive health regime that is both achievable and effective, these are some of the factors that you should pay more attention to.

THE TRUTH ABOUT DAILY MEDS

Give us this day our daily meds, and forgive us our trespasses (when we forget to take them). These days, more people take drugs than pray each day – usually on doctor's orders.

A growing number of people are taking prescription medicines as part of their normal routine. These drugs are not supposed to cure illnesses, but to prevent them. One recent survey found that in England, half of women and 43 per cent of men had taken a prescribed drug within the past week.[1] Half of those had taken more than two. Most of the drug taking is among young and otherwise healthy people: in the US, 40 per cent of regular pill poppers are under forty-four and close to half of them regard their health as excellent or very good.

On the surface, this drive to prevent disease and treat people 'just in case' looks like a good thing. It is surely better to be safe than sorry?

And the evidence shows this strategy can work. Done right, prophylaxis and prevention can stop or delay people from getting ill. But there are reasons to be cautious. Some doctors say that while preventive medicines can benefit people at high risk of developing a disease, we've gone too far. It's not unusual to see patients being prescribed ten or even fifteen different medications, often for the rest of their lives.

That raises some red flags. We don't know enough about the long-term effects of taking most drugs. And the way in which multiple medicines interact is not well understood either. Even if we know what is likely to happen if someone takes drug A, we may not know the effect of adding drug B or C (or D, E, F and so on). Drug interactions are complex and vary from person to person.

There's also the problem that drugs are being prescribed to prevent conditions they have been shown to treat, not prevent. In the absence of trials to test their preventive power, how do we know they work, or that they're not harmful when used in this way?

So should we embrace preventive drugs to promote better health, or accept treatments only when we need them? What follows is a round-up of evidence for the most common everyday medicines. It is intended to help you make informed decisions and is not definitive or individualised health advice.

Statins

They're one of the most widely prescribed medicines in the world and save many thousands of lives each year – but worries about the side effects won't go away.

Statins are thought to reduce the risk of heart attack and stroke by lowering blood cholesterol levels. One in four adult Americans over forty-five take them. However, once hailed as wonder drugs, in recent years they have been dogged by safety concerns.

In wealthier parts of the world, heart disease and strokes account for over a quarter of deaths. Many factors are at play, but conventional wisdom identifies elevated cholesterol level as one of the biggest culprits. A fatty biomolecule synthesised primarily in the liver, cholesterol is a major component of cell membranes and the myelin sheaths that insulate neurons. It is also used to make vitamin D, bile acids and steroid hormones such as cortisol and testosterone.

But it has a dark side. Excess cholesterol in the bloodstream can stick to the insides of artery walls, restricting blood flow and making clots more likely – the condition known as atherosclerosis. The relationship between diet and cholesterol is complex but that between cholesterol and heart disease is straightforward. The higher your cholesterol the more at risk you are.

Enter statins. These drugs work by inhibiting the production of an enzyme crucial for making cholesterol in the liver, and so lower blood cholesterol. The first commercial statin, Mevacor, went on sale in the US in 1987.

At first statins were prescribed to people who had already had a heart attack or stroke. But, increasingly, people with no history of heart problems were offered them too. Many trials showed that they reduced heart attacks and strokes in people with high cholesterol or who were otherwise at high risk of heart disease because of smoking, sedentariness or obesity.

One key trial known as JUPITER, published in 2008, looked at the effects of taking a statin on 17,800 people with no known history of heart disease but an elevated risk in the future. Over five years, incidence of heart attack among those who took statins more than halved.[2]

In response, the UK offered statins to anyone assessed to have at least a 20 per cent risk of developing cardiovascular disease in the next decade. That pulled seven million people in England and Wales into the statin net. As more data came in, the net was widened. In 2014 the UK threshold was lowered to 10 per cent, which pulled in an extra five million people. In the US, the threshold is even lower: 7.5 per cent risk of heart attack.

The UK's National Institute for Health and Care Excellence (NICE) estimates that its strategy should prevent 28,000 heart attacks and 16,000 strokes every year. And statins are cheaper than treating the after-effects of a heart attack or stroke.

But their increased use has met with criticism from doctors and patients suspicious of the notion of drugging people who are not unwell. For one thing, the stats mean that for every heart attack prevented, many, many more people will be taking the drug for no discernible benefit. Heart attacks may have halved in the JUPITER trial, but the absolute incidence of heart attacks in the study population was low anyway. Only ninety-nine people had a fatal heart attack during the trial period, thirty-one of whom were taking the statin. Viewed that way, less than 0.5 per cent of the people treated with the statin benefited.

Another way to look at a drug's efficacy is 'number needed to treat' (NNT), the number of people that have to be given a therapy over a specified time for one person to benefit. By this measure people with existing heart disease clearly benefit from statins: one in eighty-three people, or 1.2 per cent, have their life saved over five years. But among people without heart disease, a major heart attack is prevented for one in 140.

Admittedly, these low figures, extrapolated to whole populations, equate to hundreds of thousands of people avoiding heart attacks, but they should still give pause for thought. For one thing, prescriptions cost people money that they might be better off spending on healthier food or exercise. There is also the possibility of side effects. Patients taking statins often report muscle pain, sometimes debilitating, although this has not been reported as a major problem in the big, placebo-controlled trials.

According to a review of those clinical trials, statin therapy causes muscle pain in about fifty to a hundred patients per 10,000 treated for five years – fewer than 1 per cent.[3] But other studies have found far higher figures – typically a quarter to a third but as much as 87 per cent.[4]

There also seems to be a link to type 2 diabetes. Some research suggests that the risk rises by 10 per cent with a medium-dose statin, and then rises further in line with increased dosage. That could add up to a greater overall health burden than not taking statins.

Another criticism is that statins lull people into a false sense of security. Fixating on cholesterol diverts attention from other factors that are at least as if not more important for reducing cardiovascular risk – not smoking, eating well and exercising regularly. People on cholesterol-lowering statins frequently become more sedentary, and eat a less healthy diet, in the belief that they have solved their problem.

All of which is worth considering if you are already taking

statins, or have had them recommended to you by your doctor. There are no easy answers. But consider that NICE recommends trying lifestyle changes before resorting to the drugs.

Blood pressure medication

High blood pressure is one of the Western world's most common chronic conditions. The US spends $32 billion every year treating it – one per cent of its annual healthcare costs. Left untreated it can lead to heart attacks, strokes and kidney failure.

If lifestyle changes such as losing weight, exercising, and cutting down on salt, caffeine and alcohol don't get you out of the danger zone, doctors usually offer drugs.

The most common ones are ACE inhibitors and Angiotensin-2 receptor blockers (ARBs), both of which relax blood vessels. There are also calcium-channel blockers, which widen blood vessels, and diuretics, which flush excess water and salt out of your blood. Once you're on them, you're usually on them for life.

Many people end up taking two, three or even all four types of drugs. They are generally well tolerated though can cause various side effects including a cough, dizziness, rashes, headaches, constipation and swollen ankles. But none of these are as bad as not dealing with the problem.

Aspirin

Aspirin was first marketed in 1899, and for decades it was the world's favourite painkiller. Today, however, millions of people take it for another reason: to reduce their risk of a heart attack or stroke.

As well as being an analgesic, aspirin has powerful blood-thinning properties. It is frequently prescribed in low doses to people who have had a heart attack or stroke to protect them from having another.

It exerts this effect by inhibiting platelets, the cell fragments in the blood that help initiate clotting. In the 1980s large-scale clinical trials showed that low-dose aspirin reduces the risk of heart attacks and strokes by a third and deaths by a quarter.

But its activity on clotting is also responsible for aspirin's well-known downside, gastrointestinal irritation and bleeding, and the lesser-known risk of brain haemorrhage. Serious bleeding and even deaths do happen but are rare. Over 90 per cent of people can take low-dose aspirin without any problems. But those with a history of stomach problems, such as ulcers, or high blood pressure should consult their doctor before taking aspirin.

Some medics argue that people who have no history of heart problems should take aspirin to keep it that way. In the US, an estimated forty million adults now take preventative aspirin every day, the majority of them self-medicating.

But the US Food and Drug Administration has warned against this, judging that the risk of bleeding outweighs the benefits, even for people with a family (though not personal) history of heart disease.

UK guidelines are similar. NICE tells doctors not to prescribe aspirin or other anti-platelet meds routinely to prevent cardio-vascular disease, but says that aspirin can be considered for people at high risk, having weighed up the other risk of causing a bleed.

Even people who have been told to take aspirin by their doctor may be taking an undue risk. Out of 68,000 people in the US who had been prescribed aspirin to prevent a recurrence of heart disease, one in ten were not high enough risk to warrant it.[5]

There's another reason why people take daily aspirin. Lots of research suggests that it prevents cancer. One review of the evidence found that more than 130,000 deaths from cancer would be prevented in the UK alone if all people aged fifty to sixty-four took a low-dose aspirin every day.[6] It led to a 30 per cent reduction in both the incidence and mortality of bowel, stomach and oesophageal cancer, with smaller effects on prostate, breast and lung cancer. The researchers who led the study claim that taking low-dose aspirin is the second most powerful thing you can do to prevent cancer, after not smoking.

The effect seems to be down to aspirin's anti-inflammatory and anti-platelet properties. Platelets can shield cancerous cells in the bloodstream so they are not seen by the immune system. NICE accepts that the evidence for aspirin and bowel cancer is convincing, but as yet does not recommend taking it for that purpose.

Warfarin

Prescribed to more than ten million people in England, warfarin is a blood thinner that helps prevent clots. It is recommended for conditions in which dangerous clots can occur, including deep-vein thrombosis, heart attack and atrial fibrillation (irregular heartbeat), which is a risk factor for stroke.

Warfarin has long been used as a rat poison, albeit at doses much higher than humans take, so it is no surprise that it has side effects, including nausea and diarrhoea. There is also a risk of internal bleeding and excessive bleeding from cuts; people taking it are advised to take extra care when shaving or brushing their teeth. They are also advised to avoid alcohol and must have their dose checked frequently.

Warfarin is known to have adverse interactions with other

commonly swallowed pills, including aspirin, vitamins and other supplements, and herbal remedies.

Alternatives that don't require monitoring have been approved. However, they are expensive and don't have warfarin's long safety record.

The pill

There are millions of types of pill, but only one is known as *the* pill. That reflects the cultural and medical significance of oral contraceptives, one of the most efficient methods of contraception available. One in four women of childbearing age in the UK and the US take oral contraceptive pills, sometimes for reasons other than birth control: it also helps with menstrual pain, menstrual regulation and acne.

The pill has risks, though. A review by the European Medicines Agency concluded that some of the most popular combined contraceptive pills (which contain oestrogen and a progestin) raise the risk of deep-vein thrombosis more than previously thought.

The packaging carries a warning to that effect and doctors are told to consider patients' risk factors before prescribing. Women who are overweight, smoke or have high blood pressure may be refused. However, the danger of blood clots is still small so, on balance, the benefits of avoiding an unwanted pregnancy outweigh the risk.

The pill also slightly raises the risk of breast cancer, although in women of childbearing age the risk is very small. It also reduces the risk of other cancers – endometrial, ovarian and colorectal. Data from 46,000 women observed for up to thirty-nine years showed those who took the pill had a lower overall mortality risk, perhaps because of the cancer effect.[7]

Many women on the pill anecdotally report emotional side effects such as mood swings or low mood. Some long-term users say that when they stop taking it they feel like a different (usually better and less moody) person. But the evidence for this is woolly at best, and one recent analysis found pill users were less likely to be depressed than non-users.

Experts say that women should make their choices based on their immediate need for contraception, rather than any possible long-term health effects. If you use the pill purely for contraceptive purposes and it is causing you problems, seek medical advice about the range of other options available, from the IUS to the implant.

Hormone therapy

Few medicines have been the subject of such confusing and conflicting findings as hormone replacement therapy, which surged in popularity in the West in the 1980s and '90s but then crashed as its reputation for safety and effectiveness took a battering.

HRT does exactly what it says on the tin, replacing the hormones that decline during and after menopause and reversing many of its symptoms. Most HRT regimes involve oestrogen and progesterone (or a synthetic analogue). Testosterone is sometimes part of the mix; despite its reputation as a male sex hormone it is important in women too. Women who do not have a womb because of a hysterectomy need just oestrogen.

At first advocates claimed that, as well as relieving menopausal symptoms such as mood disorders, hot flushes and night sweats, HRT also protected women's hearts and bones and restored their libido.

That all changed in 2002 when the Women's Health Initiative, one of the biggest ever studies on HRT, showed that the treatment was not protective and raised the risk of heart disease and breast cancer. Breast cancer was already the leading cause of death among women and this extra risk scared many women and their doctors off. The number of women using it fell off a cliff, from a peak of about thirteen million in the US to about five million in 2011.

However, recent reappraisals of the trial have largely cleared HRT's name, though its reputation remains tarnished. Many of the women who took part were well over sixty, but HRT is really aimed at women in their fifties. Separating out the data on women who started HRT within ten years of becoming post-menopausal caused the breast cancer risk to drop dramatically and the heart disease risk to disappear altogether. Women who only received oestrogen had no elevated breast cancer risk. Even among women who received both hormones, the elevated breast cancer risk was less than many existing risks, such as being overweight, childless or having a pregnancy late in reproductive life.

HRT also slightly raises the risk of ovarian cancer, even if taken for just a few years, as is now the most common approach. For every 1,000 women taking HRT for five years from around age fifty, there would be one extra case of ovarian cancer.

But overall the benefits of HRT outweigh the risks. HRT is very effective against menopausal symptoms for many women as long as they start it before the age of sixty. However, the current advice is to use the smallest dose possible for the minimum time – usually no more than four years. Once stopped the cancer risks decline back to baseline. Transdermal administration via a patch or implant is safer and more effective than oral.

Whereas women have HRT, men have TRT – testosterone

replacement therapy. Millions of men use it in the belief it is an elixir of youth against the often dispiriting effects of middle age. The hormone is claimed to improve muscle strength, energy and sex drive. However, not only is there little evidence for these claims, several studies have found a link with heart disease and there are fears it may fuel cancer.

In the past, testosterone was only prescribed to men with abnormally low levels due to a congenital condition or damage to the testes. Now, though, middle-aged men are being prescribed TRT to make up for the natural decline that often comes with age.

However, the US Food and Drug Administration cautions that testosterone should only be prescribed to men with low levels caused by medical conditions. The European Medicines Agency has issued a similar statement.

The health bodies also asked manufacturers and prescribers of testosterone products to warn users about a possible risk of heart attacks and strokes after a number of studies showed an association. One trial was even terminated early due to 'an excess of cardiovascular events' among participants.

One possible mechanism for testosterone's effect on the heart could be through raising the number of red blood cells, which thickens the blood and can lead to dangerous clotting. Another concern is prostate cancer, which feeds on testosterone; drugs blocking testosterone are sometimes used to stop the cancer spreading.

Low testosterone can be a result of health problems such as obesity and diabetes, and some researchers are examining whether testosterone replacement could help. But in these cases it is often more appropriate to treat the primary condition, by losing weight, for example.

Slim, healthy older men have similar levels of testosterone to healthy young men. So the fall in testosterone with age is probably due to being fat and unfit rather than old.

THE TRUTH ABOUT MEDICAL SCREENING

In 2018, the comedian and writer Stephen Fry revealed that he had recently undergone surgery for prostate cancer. He also revealed that he had been diagnosed after a routine check-up found a high level of prostate specific antigen (PSA) – an indicator of prostate trouble but not a diagnosis of cancer – and encouraged more men to get checked out. 'I would urge any of you men of a certain age to think about getting your PSA levels checked,' he said.

It sounds like a no brainer. Common sense suggests that routine screening must be a good thing: what harm could it do systematically to test everybody for diseases such as prostate and breast cancer?

It turns out that it can do a lot of harm. As has been repeatedly demonstrated, routine screening is often, on average, worse than not screening. For every life saved through early diagnosis, many more are blighted or ended by psychological trauma, invasive investigations or unnecessary treatments. False negatives, meanwhile, can lull people who are actually ill into a false sense of security.

PSA tests are especially troublesome. Most prostate cancers aren't aggressive so don't require treatment; as the saying goes 'men usually die with prostate cancer rather than of it'. And yet the majority of men who are diagnosed via a PSA test end up having treatment with a high risk of side effects including erectile dysfunction, urinary incontinence and heart attack.

The largest ever clinical trial of PSA testing, published in the *Journal of the American Medical Association*, confirmed that while one-off tests in men with no symptoms do result in higher diagnosis, they don't increase survival rates.[8] The UK's National Health Service no longer runs a national prostate cancer screening programme because the test isn't accurate enough. You can, of course, get it done privately, but think twice before putting your hand in your pocket.

Most other mass-screening programmes have proved similarly ineffectual or even harmful and have been abandoned. Fry's exhortations probably did more harm than good. His story is inspiring but is of the anecdotal 'it worked for me' variety.

Some screening programmes save lives, however. A good guide to what works is to look at what the NHS offers, based on advice from an expert panel called the UK National Screening Committee. At present it approves routine screening for breast and bowel cancer, abdominal aortic aneurysm (AAA) screening to detect dangerous swelling of the main artery leaving the heart, and cervical smears. Some specific patient populations such as diabetics and pregnant women are offered screens too.

If private screening tempts you, consider this. Screening is still very popular in the US's privatised system. But hospitals with high uptake of CT scans – which are typically ordered to check the lungs and abdomen – perform many more kidney removal surgeries. That is because when doctors look at the images they see the kidneys too, and often stumble on innocuous tumours that they nonetheless decide to treat. This comes at a heavy price: one in fifty people who undergo the surgery die within a month, far more than would die from the tumours.

THE TRUTH ABOUT
CONSUMER GENETICS

One type of medical screening that you might be tempted to buy into is personalised genomics. Several companies claim to be able to offer you individualised health advice based on your DNA. They say they can spot genetic risk factors for many diseases including late-onset Alzheimer's, Parkinson's disease, breast cancer and type 2 diabetes.

These tests have been controversial in the past. In 2013, the US Food and Drug Administration banned the leading personal genomics company 23andMe from selling a test assessing the risk of 254 disorders. The FDA was especially concerned about false positives and negatives for breast cancer risk. A false positive might lead to a woman having unnecessary surgery and other interventions, while a false negative might lead to her ignoring actual symptoms and letting the disease progress. This is a genetic version of the screening paradox that we encountered earlier.

However, in 2017 the FDA cleared 23andMe to sell a test for ten diseases, though not including breast cancer. The diseases were chosen because there is good evidence that genetics affects risk, and that screening for them is beneficial. 23andMe had to demonstrate to the FDA that its tests could correctly identify these genetic variants with at least 99 per cent accuracy.

An extra safeguard is that customers are advised to speak to a genetic counsellor or other healthcare professional before getting the test results in order to know how to interpret them appropriately.

A major criticism of these tests is that if you are found to have a genetic risk factor, there's often little or nothing you can do about it. Some experts say you are better off not knowing, but others argue the opposite – that the information can help you plan for the future and take out the right levels of health insurance, for example.

Enter consumer epigenetics, the new frontier of personalised genomics. Companies in this area claim that their tests can deliver much more detailed and actionable health advice than is possible from a regular gene sequence.

Epigenetics refers to biochemical modifications to DNA that occur during a person's lifetime in response to lifestyle and environment. These alter the expression of genes and can have a profound effect on health. Detrimental effects can sometimes be reversed, and beneficial ones encouraged, by lifestyle changes. The companies say they will help their customers to shift their epigenomes into a healthier state.

The tests on offer include metabolic health, exposure to pollution and tobacco smoke, and levels of stress, inflammation and hypertension.

Epigenetics companies also offer tests of biological age, a scientifically well-validated measure of the amount of cellular damage incurred over the years. This is considered a better indicator of wear and tear than chronological age, and unlike chronological age, it can go down as well as up as a result of lifestyle changes (for more on this, see page 325).

THE TRUTH ABOUT
YOUR IMMUNE SYSTEM

If you're serious about not getting ill, there's one system in your body that you need to learn to look after. Your immune system is already an amazing (and amazingly complex) network of cells and molecules that detects and repels dangers of all kinds – viruses, bacteria, fungi, parasites, foreign bodies and even cancerous cells. Without it you'd die in days, and unless you have an immune deficiency of some kind – which you'd definitely know about – yours is working OK. But there are measures you can take to get it into optimal shape.

You may think that you already have a great or not so great immune system. Maybe you are one of those people who seem to catch everything doing the rounds, from coughs and colds to stomach bugs; or one of the lucky ones who never seem to get ill.

But a lot really is down to luck. There are some influences on the effectiveness of your immune system that you cannot control: your genes, your gender, your age and most importantly whether or not you have had a previous brush with a bug.

But there are plenty of things you can do something about.

Food and diet
Feed a cold, starve a fever, eat chicken soup. Food often plays a central role in folk remedies and it turns out that folk wisdom is sometimes right.

Recent research, admittedly in mice, suggests that infections caused by viruses benefit from being starved, but those caused by bacteria clear up quicker with a good feed. Viruses tend to cause respiratory tract infections including colds, while bacterial diseases are more likely to make you feverish.

Chicken soup, meanwhile, has been found to have anti-inflammatory properties and can relieve the symptoms of a cold by reducing inflammation in the nose, throat and lungs. If nothing else, it is great comfort food.

Many dietary supplements are sold as immune boosters but

only a handful actually work. The best of the lot is zinc; there is solid evidence that the supplement can shorten the duration of a cold if started within twenty-four hours of symptoms appearing. Zinc may work by stopping the cold virus from replicating or preventing it from gaining entry to cells lining the airways (for more on zinc, see page 148).

Vitamin C doesn't prevent colds, although it appears to alleviate symptoms slightly. The only other supplement with any credibility is echinacea, an extract of the purple coneflower, although again as treatment, not prevention, and the evidence is mixed.

If you want to eat your way to a better immune system the best approach is simply to eat a plentiful supply of fruit and vegetables. And exert dietary restraint. Obesity is a risk factor for infections, including respiratory, skin and urinary ones. Being bigger makes it harder to breathe, which predisposes you to colds and flu, while excess fat releases chemical signals that interfere with immune functioning.

But also be cautious about how you lose weight. Yo-yo dieting is harmful to the immune system. Frequent cycles of weight loss and gain reduce the performance of the natural killer cells that take out cancerous and virus-infected cells.

Microbiome

Antibiotics are still a highly effective treatment for most bacterial infections, but after finishing a course we are often hit by another one. That is because as well as killing the pathogens, antibiotics severely deplete the friendly bacteria living in the gut, and these are essential for a healthy immune system. Gut bacteria out-compete harmful microbes, release antimicrobial compounds and communicate with the immune system in complex ways.

It is therefore not surprising that damaging your gut flora can leave you prone to bacterial infections. So instead of razing your inner garden you should nurture it. That is the aim of pre- and probiotics, foods and supplements that are designed to boost microbiome health. Studies support the idea that they can help treat gut infections, including diarrhoea, and ward off coughs and colds.

Sleep

Even moderate sleep deprivation can put you at greater risk of catching a bug. In a classic study, the sleeping habits of healthy adults were recorded before they were exposed to a cold virus. People who slept for less than seven hours a night were almost three times as likely to catch a cold as the rest of the group. (You can read more about how to get a good night's sleep on page 292.)

Sun

There is a large class of conditions caused by immune cells attacking things they should leave well alone, namely your own body. These autoimmune disorders include type 1 diabetes, inflammatory bowel disease, multiple sclerosis and rheumatoid arthritis.

All have been linked to a lack of vitamin D, which is made by the skin when exposed to sunlight and acts as a rein on the immune system. Parts of the world with less sunlight have higher rates of autoimmune disorders.

Sunshine's effects stretch beyond those of vitamin D. Melatonin – a hormone secreted by a gland in the brain in response to changes in light – stimulates certain kinds of immune cells. (For more on vitamin D, see page 136.)

THE TRUTH ABOUT STAYING WELL

Get fit

If you are not already convinced of the need for exercise, here is yet another reason to get your trainers on: even short bursts of exercise give your immune system a kick. Adults who are physically active – doing five sessions or more of aerobic exercise a week – get minor infections half as often as people who are sedentary.

As your heart gets pumping, immune cells usually stuck in blood vessel walls are washed into the circulation, where they can do their stuff. Levels in the blood double during exercise, upping the immune system's ability to spot and respond to pathogens – what immunologists call 'immunosurveillance'.

Over-training suppresses the immune system but you have to do a hell of a lot of exercise for that to become a risk. Over-training syndrome is really only a problem for elite athletes training hard for a championship.

Age

Like most parts of the body, the immune system weakens with age. That is why older people catch more infections, are more likely to get cancer and are more prone to shingles, a painful rash caused by the chicken pox virus reactivating after lying dormant for years.

The science of anti-ageing is progressing rapidly and many of the treatments in the pipeline slow down immune senescence (for more on anti-ageing drugs, see page 320). For now the best thing to do is get fully vaccinated (a good idea at any age). As well as annual flu shots, older people can get a one-off vaccine for pneumococcal disease, which causes pneumonia and meningitis. There are also vaccines for shingles.

THE TRUTH ABOUT STRESS

Work, children, bills, groceries, chores . . . modern life can sometimes feel like a treadmill. Many of us end up frazzled, with a feeling that all this stress can't be doing us any good. But we do it anyway, perhaps buoyed up by the notion that giving in to stress is for wimps.

Big mistake. Psychological and physical stress are major triggers of disease. The thing that links them together is a process that is both vital to life and also profoundly damaging to it: inflammation.

Inflammation is the immune system's first responder. Without it, we would be at the mercy of any passing pathogen. When the body's protective barrier has been breached by injury or infection, the inflammatory response fires up, bringing redness, heat, swelling and pain.

Damaged cells secrete pro-inflammatory signalling chemicals known as cytokines, which increase blood flow to the affected area and put the rest of the immune system on high alert. The increased blood flow is what causes heat, redness, swelling and tenderness. Swelling is especially important: it is caused by blood vessels becoming more permeable, allowing white blood cells to flood into the surrounding tissue. These cells then attack and engulf invading pathogens, and later clear up the bodies.

Throughout our evolutionary history, acute inflammation has mostly worked just fine, flaring up, tackling the problem and ebbing away again when the danger has passed. But modern life has upset this delicate balance. Stress, obesity, pollution, bad diet and ageing can all tip us into a low-level state of inflammation that, rather than being confined to a specific tissue, keeps the entire body in a perpetual state of readiness for a threat that never arrives.

Persistent background inflammation does not make us ill in the short term, but in the long term it causes heart disease, type 2 diabetes and neurodegenerative disease. This 'para-inflammation' is the price we pay for our longer, calorie-rich lives.

It is also fuelled by stress. The fight or flight hormone noradrenaline, which is released in anticipation of an impending life or death situation, sets off the same chain of events as an infection or injury. In our evolutionary past stresses came and went quickly, but these days many of us are a walking time bomb of stress-induced inflammation that never ends.

Scientists have discovered, for example, that the strains of caring for a seriously ill family member increase levels of inflammatory markers in otherwise healthy people. So does a succession of short-term stresses. This is how chronic stress turns into chronic illness.

Obesity is another trigger of inflammation. A small amount of body fat is healthy, indeed necessary for regulating appetite, mood, metabolism and the immune system. But once the scales tip past 25 to 30 per cent body fat the balance shifts. (To find out how to measure your own percentage of body fat, see page 112.) Body fat stores large quantities of cytokines, and if there is too much fat on board, particularly around the organs, they can seep out, leading to chronic low-level inflammation.

High-sugar diets can also lead to gum disease. This can push the body into an inflammatory state that has been linked to an increased risk of atherosclerosis, one of the main risk factors for heart attacks and strokes.

The longer the inflammatory state persists, the more likely it is to cause problems. These can be relatively minor; for example, making it harder to shake off a minor infection such as a common cold. Or they can be life-threatening. A recent study found that inflammation is directly linked to the early stages of heart disease.[9] Over the course of three years, people with higher levels of reported stress and stress-related brain activity had higher levels of C-reactive protein, a marker of inflammation, and also had a greater risk of cardiovascular disease. When there are high levels of white blood cells in circulation, they get attracted to any fatty plaques accumulating in the arteries, making these more likely to build up and eventually rupture. This can lead the vessel wall to bleed and form a clot, which could go on to cause a heart attack or stroke.

And so we are faced with a balancing act. We need the heat of acute inflammation but don't want it to turn into an eternal flame. The simple solution is to de-stress. Find activities that give you a break from the pressures of daily life. Exercise is a great way to do it, because it gives you a fitness boost while directly battling the inflammatory monster. But de-stressing is easier said than done, so you might also need some short cuts.

It turns out that inflammation has an off switch that you can toggle. It was once thought to simply fizzle out, but it turns out that the white blood cells that initiate it also switch it off once it is no longer needed, by releasing a second set of chemicals called resolvins. Chronic inflammation is caused not just by too much 'on' signal but also by not enough 'off'.

Most modern anti-inflammatory drugs, including ibuprofen, inhibit the onset of inflammation but do nothing to bring it to an end. But an older drug, aspirin, is a different story. Low doses of aspirin not only block pro-inflammatory chemicals but also trigger the production of resolvins. Aspirin is the only drug currently available that does that – though handle it with care as it can have severe side effects (see page 238).

There are other simple things you can do. Resolvins are

synthesised from omega-3 fatty acids. So making sure you get the recommended three weekly portions of oily fish, or equivalent, might help ensure that your body has enough raw material to wind down inflammation. The link between omega-3s and the resolution of inflammation might even explain why a diet rich in these is associated with a lowered risk of heart disease.

Studies also link diets containing more fat and sugar to inflammation. This, combined with research suggesting that pigments found in fresh fruit and vegetables help to regulate inflammation, is another reason to watch what you eat.

So, yes, the answers to stress are familiar ones. Eat well, exercise, lose weight. These work not only by reducing inflammation but, paradoxically, by exerting a different kind of stress on your body. A little bit of stress is good for you. This is due to a biological process called hormesis, whereby exposure to low levels of stress or toxins makes you more resilient to greater challenges.

The defining characteristic of hormesis is the 'biphasic dose response', in which high doses of a substance are toxic but low doses are beneficial. The aphorism 'what doesn't kill you makes you stronger' turns out to be true.

In fact, a pretty potent elixir of youth might look like this: an X-ray, mild starvation, a couple of beers and a dose of heatstroke. If that sounds like a recipe for the exact opposite of longevity, think again. Hormesis has been shown to extend longevity in yeast, fruit flies, protozoans, worms and rodents. If the findings extend to people, it could stretch the average healthy human lifespan to ninety.

There is some evidence that hormesis has positive effects on human longevity. Between 1980 and 1988, researchers in the

US tracked 28,000 nuclear shipyard workers to study the effects of low doses of radiation. They found that the mortality rate of these workers was 24 per cent lower than in a control group of 32,500 shipyard workers of similar ages who were not exposed to radiation. An earlier study found similar low death rates among radiologists compared with other doctors.

How so? Stressors seem to kick-start natural repair mechanisms. One is heat-shock proteins, which are produced when cells are exposed to high temperatures, toxins or inflammation. They protect other proteins from damage by binding to them and shielding them from attack. Another, sirtuin 1, senses cellular stress and activates genes that code for protective proteins such as antioxidants and cell-membrane stabilisers. DNA repair enzymes kick in too.

If the damage is not too severe, these repair systems over-compensate, building up enough oomph to repair unrelated damage as well. Given that damage equals ageing, this is nothing less than rejuvenation.

You may not even have to expose yourself to poisonous chemicals or radiation to see the benefits of hormesis. Caloric restriction – the near-starvation diet that is the only reliable way so far of increasing lifespan in animals – probably works because it is a low-level stressor.

These molecular defences evolved to protect us from naturally occurring threats, but there are ways to activate them without stressing yourself. One is eating lots of fruits and vegetables. There is much evidence that a diet rich in plants reduces your risk of cardiovascular disease, cancer and some neurodegenerative disorders. The standard explanation for this is that fruit and vegetables contain high levels of antioxidants, such as carotenoids and flavonoids, which neutralise free radicals.

However, most plant antioxidants only mop up free radicals at high concentrations that cannot be achieved by eating normal amounts of fruit and vegetables. Clinical trials of high-dose antioxidants have failed to show that they prevent or treat these diseases.

Hormesis provides a different explanation. Antioxidants are part of a wider class of plant chemicals called phytochemicals that are toxic at high doses but good for us at lower doses. They probably evolved as natural pesticides. The amounts we eat are insufficient to be toxic to the human body, but are enough to activate our molecular stress responses. In other words, they are hormetic stressors.

For example, resveratrol, the chemical supposedly responsible for the health benefits of red wine, activates sirtuin 1; while sulforaphane from broccoli activates a protein called Nrf2, which switches on genes for antioxidant and detoxification enzymes. Nrf2 is also activated by curcumin from turmeric. Allicin from garlic and capsaicin from chilli also induce a mild stress response.

Many other healthy compounds in our diets – including vitamin A, vitamin B6, selenium, iron and zinc – are toxic at high doses and may exert some of their low-dose effects via hormesis.

Hormesis is also at least partly responsible for the health benefits of exercise and dieting. Increasing energy use and cutting down on calories induce a state called 'mild metabolic stress'. Cells respond to this by activating stress-response pathways that increase their ability to take up glucose in response to insulin.

Mild metabolic stress also causes cells in the heart and gut to produce proteins that decrease heart rate and blood pressure and increase gut motility, reducing the risk of heart disease, stroke and colon cancer. Hormeses may thus be behind the beneficial effects of fasting.

So maybe it's not about being 'Zen' all the time. In the right doses, stress can be a tonic. Just don't let it get out of hand.

THE TRUTH ABOUT BEING AT ONE WITH NATURE

Doctors in the Shetland Islands have begun prescribing some pretty wild treatments. Alongside conventional medicine, people with a range of physical and mental illnesses – including heart disease and diabetes – are being urged to take up hill walking, beachcombing, bird-watching or even cloud spotting. The goal is not to get them exercising but to immerse them in nature.

A similar scheme in New Zealand found that six to eight months after receiving a 'nature prescription', two-thirds of patients were more active and felt healthier, and almost half had lost weight.

Japan and China, meanwhile, have a tradition of Shinrin-Yoku, or forest bathing, which involves walking through a forest breathing slowly while consciously appreciating the sights, sounds, smells and feel of nature. It sounds like New Age nonsense – it is literally tree hugging – but has been shown to have significant cardiovascular and immune system benefits.

It has long been known that people living in greener neighbourhoods tend to have better cardiovascular health and lower levels of stress, regardless of how wealthy they are. Recent research also suggests that city dwellers living near green spaces are at lower risk of type 2 diabetes than those without access.[10]

Connecting with nature is good for us – mostly for mental health but also for physical well-being. Heart rate slows down, we breathe better and our immune responses are improved. Exactly why is not completely understood. Various explanations have been proposed, many related to a reduced stress response. But being outdoors boosts your exposure to bright light, which has a number of health benefits (for more on this, see page 308). The air is often fresher. There are the benefits of exercise, social contact and time away from workaday worries. Some even believe that we have an evolved an urge to seek out our natural habitat – trees, grassland and sky – and respond positively to it.

The amount of time needed to spend in nature to reap its benefits turns out to be two hours a week. That was the key finding from a recent survey of 20,000 people in the UK, who recorded how much time they spent in green spaces – including city parks – over the course of a week, plus details of their health and well-being.[11] The researchers controlled for the possibility that the health benefits might be a product of physical activity rather than contact with nature.

A two-hour nature bath can be taken in one go or be spread out over the week. For most people it wouldn't take much effort to hit the target. The survey also revealed that the average amount of time people that people spend in natural environments each week is already ninety-four minutes.

Consider two hours to be a minimum. Stay longer and the benefits continue to accrue, albeit at a diminishing rate. After five hours, though, the benefits have maxed out and you can scuttle back indoors.

You don't have to be moving through nature; just sitting or standing will do. But if you do exercise, why not kill two birds with one stone and do your workout in the park or woods? Failing that, adjust your routine so that it incorporates a bit of nature bathing – a walk through a park on the way to work, for example.

If you really cannot get out into a green space, it turns out that looking at pictures of natural scenes or listening to recordings of natural sounds such as birdsong replicate some of the benefits, reducing people's heart rates and blood pressure. You can also simulate the smell of a forest: diffusing pine oils in bedrooms as people slept led them to up their production of immune cells called natural killer cells, which fight viruses and cancers. So even if your nature bathing is thoroughly unnatural, there's no excuse for missing out on the benefits.

If you literally want to bathe in nature, there's the increasingly popular pastime of wild swimming. Immersing oneself (*sans* wetsuit) in a lake, pond or sea – especially a cold one – is claimed to have numerous health benefits, including boosting the immune system, upping metabolic rate and stimulating the natural antioxidants that protect us against free radical damage. Anecdotally, people who go winter swimming – defined as when both the air and water temperature are below 18 °C – say it makes them feel invigorated.

The evidence for this is watery, however: any benefits are hard to tease out from the general benefits of exercise, and a study of regular winter swimmers in San Francisco Bay saw no reduction in upper respiratory tract infections. Regular winter swimmers do become physiologically adapted, shivering less and becoming more tolerant of cold, but whether they have general health benefits is not known.[12] There are dangers too. Swimming in wild waters is hazardous for all sorts of reasons and sudden immersion in cold water can cause heart attacks and strokes; about 1,000 people die from cold water-induced cardiovascular events in the UK each year. However, a dip in a pond or lake in summer probably won't do you any harm. If nothing else, it contributes to your two hours a week quota of nature bathing, and we know for sure that is good for you.

THE TRUTH ABOUT AIR POLLUTION

Around the world, one in nine early deaths is caused by air pollution, according to the WHO. That may conjure up images of the brown smog that hovers over so many Asian cities, but air pollution is a major health concern in rich countries too. It is difficult to put a number on it, but toxic air is estimated to cause 200,000 untimely deaths each year in the US and nearly half a million in Europe, with tens of thousands of those in the UK.

The problem is that the air doesn't have to look toxic to be toxic. The air pollution that blights cities in Europe and North America is largely invisible. If you travel on, work or live near a busy road, you're breathing it in, even when the sky is a brilliant blue. That killer is the gas nitrogen dioxide (NO_2), and Western cities have some of the world's highest levels. Diesel vehicles are the biggest source, and Europe's roads are full of such cars thanks to well-intentioned but ultimately misguided green policies that encouraged people to buy them.

As well as NO_2, city dwellers are exposed to high levels of soot and other forms of particulate pollution, which cause different but even deadlier health problems, including lung cancer. Not all of this comes from fumes. In cities, about half the particulates come from brake pads and a tenth from tyres. Another quarter simply come from traffic stirring up dirt.

Air pollution doesn't kill directly, but it aggravates other things that are likely to kill and so takes years off people's lives.

Epidemiologists measure its aggregate toll in terms of equivalent lives lost per year. A recent report concluded that particulate pollution causes 30,000 of these 'deaths' in the UK, and NO_2 causes 10,000.[13] That means air pollution kills more people than obesity or alcohol – only smoking is more dangerous.

Toxicologists used to debate whether NO_2 was harmful itself, or merely seemed so because high levels of it generally coincided with high levels of particulate pollutants. Now there is strong evidence that NO_2 has a wide range of direct harmful effects, including increasing the risk of respiratory infections and cardiovascular disease.

Air pollution is also suspected to increase the risk of conditions including diabetes, dementia and autism. A recent study found millions of tiny iron particles in brain tissues, probably from exhaust fumes.[14] Toxicologists say there is no safe limit for air pollution. Even tiny amounts start to damage your health.

There are limits intended to protect city dwellers' health somewhat, but they are regularly broken. EU regulations say that NO_2 should average no more than forty micrograms per cubic metre over a year (the US standard is less strict: 100 μg/m3 over a year), and should not exceed an average 200 μg/m3 for over an hour more than eighteen times in a year. These limits are regularly breached in half of EU countries.

In London, the city centre invariably exceeds the 40 μg/m3 annual average limit, and in some parts the eighteen times in a year rule is usually exceeded by the end of January. All of central London also exceeds the EU limit for particulates smaller than ten micrometres in diameter (PM10s), which in any case is twice that recommended by the WHO.

We still have much to learn about how air pollution causes ill health. Its biological effects are likely to be multiple, complex and interdependent. Inflammation is likely to be one major factor (for more on this, see page 252). Studies suggest PM2.5s (particulates smaller than 2.5 micrometres in diameter), NO_2 and ozone trigger it.

One way to get a handle on this is to look for correlations between exposure to pollution and certain diseases. For instance, elevated PM2.5 exposure is associated with an increase in heart attacks and lung cancer. One study found that people living within fifty metres of a major road are 7 per cent more likely to develop dementia than those 300 metres or more away.[15] Other research has linked air pollution with diabetes, kidney diseases, Alzheimer's, premature births and mental illness.

The good news is that pollution levels are declining, albeit slowly, and there are things you can do to protect yourself. One way is to avoid travelling in cars. You may assume that being inside the car rather than outside on the road shields you from pollution, but you'd be wrong. Car passengers in busy traffic are exposed to twenty-five times more carbon monoxide, and between two and three times more PM2.5s, than pedestrians and cyclists on the same road. One reason may be that most cars have an air intake at the front, directly behind the exhaust pipe of the vehicle in front. Air filters in most modern cars block large particles like dust and pollen, but are unlikely to capture all fine particulates and do nothing to stop NO_2 and ultra-fine particles. Switching to a vehicle's recirculation mode, where cooling air is reused rather than sucked in from outside, can help. The risks are lower in free-flowing traffic but car passengers are still worse off than cyclists and pedestrians. Bus passengers are less exposed than car passengers.

People moving under their own steam get more exercise and are not contributing to the pollution. The positive effects of exercise hugely outweigh the damaging effects of the pollution, especially if you cycle or walk on quieter roads. There is less pollution to start with on back streets and they are often shielded from traffic by buildings. But if you can't avoid the busy road, consider that concentrations of nitrogen oxides (NO_x) and PM2.5s from cars fall exponentially with distance from the exhaust pipe; even walking on the inner side of the pavement rather than right next to the traffic can make a significant difference, as can walking on the side of the road where the wind is blowing from.

Living out of the city generally reduces exposure to NO_2, carbon monoxide and particulates, although ozone levels are on average higher in rural areas. Overall, country air is less polluted.

Face masks can help, at least with particulates. Choose one with an N95 rating, which means it has been certified by the US National Institute for Occupational Safety and Health to filter out 95 per cent of airborne particles larger than 0.3 μm.

Cycling masks range from 55 to 85 on the same scale; those with exhalation valves may work better. Professional surgical masks are surprisingly good, rating as high as 80. Cotton masks are generally quite ineffective. Masks don't keep out NO_2 unless they have a charcoal filter. The key attribute for all masks is a snug fit: if a mask doesn't perfectly follow the contours of your face then you're not really wearing a mask.

Infants and small children are at greater risk because they are nearer exhaust level, either from being in a pushchair or by simply being short. Ditto riders of recumbent bikes. Experts advise parents to use pram covers, especially near busy traffic.

Indoor air pollution is also being recognised as a health hazard. Open a window when cooking on gas as burning natural gas (methane) produces NO_2. In fact, burning anything in your home produces harmful pollutants, so give up on the candles, joss sticks, open fires and log burners.

In winter, modern log-burning stoves are responsible for a huge amount of air pollution in northern cities such as London and Copenhagen. Wood smoke contains many of the same nasties as tobacco smoke.

Houseplants are said to mop up indoor pollution but only make a tiny dent.

What are you breathing in?
PARTICULATE MATTER (PM)
Sources: car exhaust, power plants, factories, gas cookers, open fires, log burners, volcanoes, dust storms, forest fires.

Health effects: Any dust or droplet less than 10 micrometres (μm) across (called PM10) can penetrate deep into your lungs. Those smaller than 2.5 μm (PM2.5) are the most damaging air pollutants. They include ultra-fine particles, which are smaller than 0.1 μm. Long-term exposure to PM2.5s can impair lung and heart function and increase mortality, especially among those at higher risk of heart disease and stroke.

NITROGEN OXIDES (NO_x)
Sources: road transport – especially diesel engines – as well as indoor heating and power stations.

Health effects: NO_2 is seen as the second most harmful pollutant after PM2.5. Exposure can trigger respiratory problems and inflammation, though long-term effects are unclear.

GROUND-LEVEL OZONE (O_3)
Sources: reactions between other chemicals including NO_x and volatile organic compounds, especially on warm, sunny days.

Health effects: Ozone can cause wheezing, shortness of breath, inflamed and damaged airways and asthma attacks. O_3 is a powerful oxidant, so inflicts oxidative damage on cells and tissues.

THE TRUTH ABOUT LONELINESS

One of the most underreported and neglected health problems in the Western world is loneliness. Surveys suggest that it has reached epidemic levels. In the UK, for example, more than nine million adults report being lonely often or always. Children and people with disabilities report especially high levels of loneliness.

Loneliness is often assumed to be a problem that predominantly affects the elderly or vulnerable. There is some truth to that: nearly half of people of retirement age say the television is their main source of company. But loneliness can and does affect anybody.

This adds up to an epidemic – and that is not a word used lightly. Chronic loneliness is extremely damaging to people's health. It is largely seen as a psychological problem, but its effects are also physical. Left unchecked, loneliness can have a health impact as detrimental as smoking or obesity.

It is not hard to find reasons. More people than ever live alone, and the number of single-parent households is rising. The demands of education and work mean that many of us live far away from families and friends. At the same time, technology has changed the way we work, shop, socialise and entertain ourselves, largely serving to reduce the amount of face-to-face contact we get.

There are life events that make us especially vulnerable to loneliness, such as moving home; changing job; bereavement;

divorce or separation; having a new baby or having an older child leave home. But these periods of acute loneliness can and often do pass. What is more damaging is chronic loneliness.

If you feel lonely, you probably are. Loneliness is a highly subjective phenomenon, and we all need different amounts of social contact. Some people can spend much of their time alone but suffer no loneliness, while others get lonely in a crowd. It is this, rather than the objective amount of social contact we have, that determines how lonely we are. The differences are partly innate: studies of twins suggest that genes predispose some people to a greater need for social interaction.

Lonely people are at increased risk of just about every major chronic illness going: cardiovascular disease, neurodegenerative diseases, cancer. A meta-analysis of nearly 150 studies found that a lack of social interaction had the same negative effect on risk of death as smoking, alcohol, lack of exercise and obesity.[16] Loneliness increases the odds of early mortality by 26 per cent, which is about the same as obesity.

This may be in part because it interferes with sleep. People who rate themselves as lonely are more likely to sleep fitfully, feel tired during the day and have trouble concentrating; sleep deprivation is a known risk factor for many chronic illnesses.

But that is not the whole story. Loneliness increases the risk of mental health problems such as anxiety, stress, depression and eating disorders, which can have a knock-on effect on physical health.[17] It also lowers willpower (for more on this, see page 218), so lonely people are more likely to eat unhealthily or fail to exercise.

Perhaps the biggest effect is on the immune system. One study compared inflammation in persistently lonely versus socially active middle-aged adults.[18] In the lonelier group, the activity of genes responsible for inflammation was turned right up. Inflammation is the body's first line of defence against injury and bacterial infection, but too much inflammation has been linked to cancer, depression, Alzheimer's disease and obesity.

If loneliness is so bad for us, why have we evolved to feel it? The answer lies in our evolutionary past. Humans are an obligatorily gregarious species, which means that our survival and reproductive success are dependent on cooperation with others. Whether finding food or fending off an attack by predators or enemy tribes, there is safety in numbers.

This evolutionary interpretation helps explain the immune system changes. Ramping up inflammation and dampening other immune function is what's called the conserved transcriptional response to adversity. It usually happens at times of heightened danger, priming the immune system to be more effective at dealing with injuries and the bacterial infections that follow.

Once that acute terror has passed, the system would normally shift back to normal. But chronically lonely people feel a constant mortal threat because of their lack of backup. Their bodies are ceaselessly primed for an injury that never comes.

This heightened sense of threat may also be why loneliness is the enemy of sleep. If you are alone you cannot afford to lose consciousness, and your brain responds accordingly.

So if you are lonely, what can you do? One thing to be aware of is that loneliness can be a vicious circle. Inflammation dampens down brain areas that motivate you to interact with others. This probably evolved to encourage sick people to put themselves in a kind of self-imposed quarantine. But it can make us worse at reading social situations and prevent lonely people from seeking out the company they desperately need. So it is easy to see why, left unchecked for too long, loneliness can spiral out of control.

If there's one factor that stands out in alleviating loneliness, then it is the quality, rather than quantity, of relationships. This is also rooted in our evolutionary past, when having a small number of intensely close family or friends was key to survival. This inner circle typically numbered about five people.

To maintain those crucial relationships, there's an easy formula – you need to dedicate 40 per cent of your total social time to them. That usually means making more of an effort to socialise with them rather than, say, just live in the same house. But beware of social media. All too often it becomes antisocial media. Though it ought to be an antidote to loneliness, helping people to connect even when they can't meet up in person, research suggests that heavier users tend to be lonelier.

Is that because social media causes loneliness, or because lonely people use social media more? The jury is still out on that one, but in the meantime there are some tips to prevent social media from making you lonely.

What seems to be particularly damaging is passive use, or lurking, when people read but don't post or interact. This is probably bad because people curate their social media output to make their lives seem more fun and interesting than they are, provoking feelings of isolation in people who are vulnerable to loneliness or already feeling isolated. Other small changes, such as culling distant acquaintances and setting notifications only for updates from real friends, can help too.

It sounds like hard work but it pays off. Being socially active is one of the best ways to add years to your life. Good relationships with family, friends, neighbours and even pets all do the trick, but the biggest longevity boost comes from marriage or an equivalent life partner relationship. The effect was first noted in the nineteenth century, and studies carried out since then suggest that marriage adds as much as seven years to your life. A recent study of long-lived families in Denmark found unusually low rates of divorce.[19] Having lots of siblings also seems to help. The Danish study also found that long-lived families are usually larger than average.

Exactly how close relationships promote longevity is not entirely clear, but is probably a complex mixture of socio-economic, psychological and physiological factors. For example, people in supportive relationships may handle stress better.

There is a flip side, however. People who lose a life partner are more likely than average to die in the next two years. Like marriage, long life is a joint effort.

To measure loneliness, researchers commonly use the UCLA Loneliness Scale.[20] The full version is twenty questions long, but here is a shorter one.

Answer the questions using a scale from 1 to 4, where 1 = never, 2 = rarely, 3 = sometimes and 4 = always, then calculate your total score.

1. How often do you feel unhappy doing so many things alone?
2. How often do you feel you have no one to talk to?
3. How often do you feel you cannot tolerate being so alone?
4. How often do you feel as if no one understands you?
5. How often do you find yourself waiting for people to call or write?
6. How often do you feel completely alone?
7. How often do you feel unable to reach out and communicate with those around you?
8. How often do you feel starved of company?

9. How often do you feel it is difficult for you to make friends?

10. How often do you feel shut out and excluded by others?

How you scored:

- **20** is the average score on this survey
- **25-29** points to a high level of loneliness
- **30 +** points to a very high level of loneliness

THE TRUTH ABOUT SUNSHINE

Here comes the sun – and here comes the sunscreen. Everyone knows that too much sun equals skin cancer. So slap on the cream and stick on a hat and, whatever you do, don't let yourself burn.

There's a very good reason for this advice. The UV in sunlight has a very dark side. It causes damage to the skin that can develop into cancer, especially malignant melanoma, the deadliest form of skin cancer. In fact, sunlight exposure is by far the main cause of melanoma. Sunburn is a painful warning that you have overdone it. The relationship between sunburn and melanoma is complex, but each time you get burned you have elevated your risk of melanoma in the future. In fact, all sun exposure causes cell mutations that may eventually lead to cancer.

Advice to wear sunscreen seems to be achieving its aim. In Australia, the skin cancer capital of the world, rates of melanoma are now going down after increasing for decades.

At the same time, however, doctors are becoming increasingly worried that the advice has gone too far. All over the world, vitamin D deficiency is increasing, and advice to avoid the sun or wear sunscreen has taken much of the blame. Sun striking the skin stimulates it to produce vitamin D; avoiding the sun or screening it out with sun cream vastly reduces how much vitamin D your body makes, leaving it reliant on diet or supplements, neither of which deliver enough.

Around a billion people are vitamin D deficient worldwide, and health conditions caused by it are on the rise. These include osteomalacia, or softening of the bones. In childhood this is known as rickets.

Low levels of vitamin D are also associated with weaker teeth, infections, cardiovascular problems and autoimmune and inflammatory diseases such as multiple sclerosis. And although vitamin D supplements are touted as an alternative, they don't seem to work nearly as well as sunshine.

If anything, the problem is getting worse. There is a trend towards greater sun protection factors – SPF100 is now available – and most Westerners spend the majority of their lives indoors. Meanwhile, moisturisers and cosmetics now routinely contain sunscreen.

About 10 per cent of people in the UK have insufficient levels of vitamin D during the summer, rising to nearly 40 per cent during the winter. For this reason, the UK's Scientific Advisory Committee on Nutrition recommends that everyone should consider taking vitamin D supplements during winter.

Modern medical interest in sunlight dawned in the early twentieth century, following observations that lack of light was associated with rickets. By the 1920s, sunlight was being touted as a cure for pretty much every illness under the sun.

Scientists soon identified one of the key mechanisms by which sunlight promotes health. When the ultraviolet (UV) B rays in sunlight hit the skin, they spur the conversion of cholesterol into vitamin D3, an inactive precursor. This enters the bloodstream and is further metabolised into the active form of vitamin D elsewhere in the body. At least 80 per cent and sometimes all of the vitamin D in our bodies comes from exposing the skin to UVB. Excess is stored in fat cells, and you can build up a large enough stockpile to see you through several months of producing none.

By definition, vitamins are vital to health and deficiencies lead to disease or even death. Vitamin D is needed by bone and muscle cells to keep them strong and healthy, and it is also important for certain immune cells.

Vitamin D isn't the only way sunlight affects our health. UV light striking the skin may also help keep our immune systems in good working order. The outer layer of skin contains cells called keratinocytes, which convert UV into signals that keep the immune system in check, damping it down. This may be essential for preventing autoimmune diseases. However, sunlight's effect on immune suppression also has a downside. With a less active immune system, cancers can grow more easily.

Nonetheless, people with high sun exposures have higher life expectancies, on average, than sun avoiders, despite facing an increased risk of cancer. A large twenty-year study in Sweden found that, on average, women who spent more time in the sun lived one to two years longer than sun avoiders, even after adjusting for factors such as income, education and exercise.[21] This reduced life expectancy among sun avoiders was mostly due to a greater risk of death from cardiovascular disease, type 2 diabetes, autoimmune disease or chronic lung disease.

Sunlight also stimulates the production and storage of nitric oxide, a potent dilator of blood vessels that reduces blood pressure. This may be why people's blood pressure readings are lower in summer than in winter, and why cardiovascular disease is more prevalent at higher latitudes. If you expose somebody to the equivalent of about twenty minutes of summer sunlight, their blood pressure drops.[22]

This sun-fuelled nitric oxide may have other functions. Mice fed a high-fat diet are protected from weight gain and metabolic diseases if they get regular exposure to UV light, but not if nitric oxide production is blocked. Nitric oxide also promoted wound healing in mice, which may be why being in the sun can help with that.

Low exposure to daylight has also been blamed for the global rise in short-sightedness over the past sixty years. In 1950s China, only 15 per cent of people were myopic. Today around three-quarters of young adults living in East Asia are myopic. By 2050, half of the world's population could be myopic. Spending just two hours a day outdoors in sunlight could prevent myopia.

On top of that, about 1.7 billion people have a latent tuber-culosis infection, which is kept in check by the immune system. Vitamin D deficiency has been linked to it breaking out. In the UK, TB incidence peaks in the summer, which has been asso-ciated with dips in sunshine six months earlier.

We are therefore in a bit of a bind. Sunlight is both good for us and bad for us. How can we balance the risks to ensure we get the right amount of sunshine?

Most dermatologists still believe that the skin cancer risk outweighs that of vitamin D deficiency. For example, getting sunburnt just once every two years triples the risk of melanoma.[23]

Australia's health authorities, who know a thing or two about skin cancer, advise people to consult the UV index, a measure of how strong the sun's UV rays are on a linear scale from 1 to 12. The index is designed to measure sunburn risk. Australians are advised to take evasive action when the index is just 3 or above. To put that in perspective, the average midday UV index in London in April is 4; in July it is 6 and in January 1. It almost never exceeds 8 in the UK, but sunnier places regularly go much higher and even bust the supposed maximum, 12. During autumn and winter, though, people living in southern Australia, where vitamin D deficiency is more of a risk, are encouraged to head outside in the middle of the day with some skin uncovered.

That won't work in countries at higher latitudes, such as the UK, because the sun doesn't rise high enough during winter for the UVB rays to reach the ground. In these countries – roughly anywhere north of San Francisco or southern Spain, or south of Melbourne, Australia – winter residents are dependent on the vitamin D stockpiled during warmer months, supplemented by that obtained from our diets. Foods such as oily fish, egg yolks, liver and some fortified breakfast cereals are rich sources, but we don't eat anywhere near enough of these to fill our boots.

Many people also take vitamin D tablets, but according to a recent review, vitamin D supplementation has only two clear benefits: it prevents upper respiratory tract infections and stops existing asthma from getting worse.[24] What is more, the form derived from sunlight is stored in the body for twice as long as that from supplements.

To make vitamin D the natural way, you need exposure to UVB rays, which peak around noon. You make relatively little vitamin D in the morning and late afternoon when the sun is lower in the sky. You don't need to spend hours in the summer sun to ensure you build up adequate vitamin D for the year. The minimal dose of sunlight varies, but it is lower than the dose that will give you sunburn. During a typical British summer, as little as five minutes exposure a day is ample. The best thing to do is to catch a bit of time outdoors around noon. You can't make vitamin D by sitting next to a closed window, though, because UVB rays don't penetrate glass. You can burn, however, because some UVA rays can get through.

You still make vitamin D while wearing sunscreen, though at a slower rate. It is called sunscreen for a reason: it screens, but doesn't block, the sun. Whatever the SPF, some UV will get through.

SPF30 allows 3.3 per cent of UV through; SPF50 2 per cent, and SPF100 1 per cent. And that assumes you are applying it properly, which means using roughly thirty-five millilitres per application. That's a teaspoon for each limb, another teaspoon for your front, another for your back and yet another for the face and neck, reapplying every two hours. Most people apply less than this amount, and don't reapply often enough.

On top of that, the SPF rating is primarily a measure of protection against UVB rays, but UVA rays also damage skin. For UVA protection, look for sunscreens labelled 'broad spectrum protection' with a symbol of UVA in a circle or with a high-UVA star rating.

Also consider your skin type. People with darker skin will need to spend longer in the sun to generate vitamin D and nitric oxide, although they also take longer to burn.

A few years ago I went to see my doctor about a persistent fungal infection on my toenails. I expected it to be a quick visit, but I was wrong. The doctor noted my age – forty-five at the time – and the fact that I hadn't been to see him for a while. He asked some lifestyle questions, weighed me, checked my heart rate and took my blood pressure, which he said was far too high. He gave me an appointment to return the next week, this time for blood tests.

It turned out that I was generally fine – cholesterol levels OK, no sign of prostate problems, a BMI shading into the overweight category and liver enzymes a bit elevated but nothing to worry about if I cut down on alcohol. But my blood pressure was through the roof. I had hypertension. Over the next few weeks I had further tests and examinations, which I consistently flunked. In the end the doctor prescribed me some blood pressure pills, which lowered it a bit but not enough. He doubled the dose, and my hypertension is now under control.

At the time I was mortified – here I was in early middle age and I was already on old person medication for the rest of my life. But I'm now relieved. My blood pressure was in dangerous territory, and my risk of a cardiovascular accident much higher than it should have been at my age. My dad has high blood pressure and so did his dad, and he died of heart failure aged just sixty-nine. Fingers crossed I'll make it way past that, at least due in part to my doctor's vigilance.

For me, prevention is better than cure. And as we've seen in this chapter, that covers a multitude of prevention strategies, from medication to avoiding some of the most insidious health hazards of modern life. If you haven't been to see your doctor for a few years, go. You may end up getting more than you bargained for, in the form of a prescription for something you didn't even know you had. But you'd rather know.

And if you are worried about my toenails, they are fine now, thanks.

THE TRUTH
ABOUT SLEEP

WITHIN A FEW hours of reading this you will lose consciousness and slip into a strange twilight world. Where does your mind go during that altered state – or more accurately states – we call sleep? And what is so vital about it that we must spend a third of our lives sleeping?

Until relatively recently, scientists were rather uninterested in sleep. But in 1953, the year that Everest was conquered and DNA's structure discovered, a PhD student called Eugene Aserinsky did what no one had thought to do before. He spent hours staring at the eyelids of a sleeping person. What he saw was amazing. Instead of the slow periodic motion he'd been expecting, he saw frantic, jerky eye movements.

Until then, sleep had been seen as a pretty passive and uninteresting state. But Aserinsky and his PhD adviser, physiologist Nathaniel Kleitman of the University of Chicago, went on to show that this 'rapid eye movement' was correlated with massively increased brain activity and dreaming. Their discovery is widely regarded as having kick-started the modern discipline of sleep science.

Research has mostly focused on the brain, but in recent years a second revolution has taken place in sleep science. We now know that sleep has a profound effect on general health, physical as well as mental.

But sleeping well is maddeningly, frustratingly difficult. Sleep is one of life's pleasures but also a constant source of anxiety and misery. We do it every day, but many of us don't feel like we've got the hang of it. Why is sleep so hard to do well? How much is enough? Can you get away with less? It's enough to keep you awake at night. Fortunately, there are ways you can learn to sleep the good sleep.

THE TRUTH ABOUT WHAT HAPPENS WHEN WE SLEEP

Sleep is not one thing but many; when we are asleep we cycle through different phases. These fall into two broad categories – REM sleep and non-REM (NREM) sleep. REM stands for rapid eye movement, after its most obvious outward manifest-ation. REM is also called dreaming sleep, and it is when the majority of dreaming occurs. But dreams also happen in deep sleep, though they tend to be less vivid, less emotionally charged and less memorable.

NREM sleep is divided into three different stages. Stage 1 is light sleep, which occurs just after you doze off and before you wake up. It lasts just a few minutes and is a mysterious border-land between wakefulness and sleep. If roused from stage 1 sleep, people often think they were already awake.

Stage 1 sleep sometimes features bizarre sensory experiences such as hallucinations and lucid dreaming, collectively known as hypnagogia. It is also when hypnagogic jerks happen. These are the lurches or twitches you can experience as you drift off, often accompanied by the sensation of falling. The cause remains a mystery. One idea is that you start dreaming before you are fully asleep. Another is that the twitches are a by-product of your nervous system relaxing as you drift off. A bizarre event that can occur during stage 1 is exploding head syndrome, which is the sensation of a loud noise like an exploding bomb or a gunshot. It affects about one in ten of us and it tends to start around the age of fifty. Nobody knows what causes it. Despite its name, the condition is harmless.

From stage 1 you go into stage 2. This is deeper and less eventful than stage 1, with slower brainwaves interrupted by brief bursts of electrical activity and no dreaming. It lasts about twenty-five minutes before you transition into stage 3, also called deep or slow-wave sleep.

Stage 3 is the deepest and most restorative phase of sleep. Your heart slows down and brainwaves become long and regular. It is hard to wake somebody up from this state and it can take them up to an hour to shake off the grogginess – or sleep inertia – that follows. Some dreaming occurs. Stage 3 lasts up to forty minutes, then you cycle back up through stage 2 and into REM.

This is when most dreaming occurs, and your muscles are paralysed during it to stop you from acting out your dreams. REM is thought to be when your brain processes memories and emotions but, like the rest of sleep, its function remains a matter of debate. Brain scans done during REM are very similar to those of fully conscious wakefulness. The first bout of REM lasts a few minutes but each successive bout lasts longer. A full eight-hour sleep features about two hours of REM, with the bulk crammed into the end of the night.

After every round of REM you go back into stage 1, and then go through the phases again. A typical night's sleep involves cycling through these stages several times. If you do wake up in the night it will probably be after a bout of REM, during the next instalment of stage 1.

The need to sleep is controlled by a two-tier system. The first tier is our circadian clock, which controls the twenty-four-hour cycle of body temperature, blood pressure and release of hormones. It makes us feel sleepy around midnight, but also – to a lesser extent – mid-afternoon (if you've ever felt the post-lunch slump, you can blame your biology). It relies on light to keep your sleep/wake pattern to roughly twenty-four hours. We'll look at what happens when this is disrupted on page 305.

Then there's sleep drive or sleep pressure. Throughout your waking hours a chemical called adenosine builds up in your brain, sending signals that increase your desire for sleep. After about sixteen hours, resisting that pressure becomes extremely difficult. Caffeine keeps you awake by blocking adenosine receptors in the brain.

THE TRUTH ABOUT WHY WE SLEEP

On the face of it the answer seems obvious: we sleep so that our brains and bodies can rest and repair themselves. But why do these require us to become unconscious? Why not do it while conscious, so that we can watch out for threats?

Sleep is such a universal phenomenon that it must be doing something useful. All animals sleep; even tiny flies have periods of inactivity from which they are not easy to rouse, suggesting sleep is a requirement of the simplest of animals. But surveying the animal kingdom reveals no clear correlation between sleep habits and anything at all. There is bewildering diversity.

Some bats sleep for twenty hours a day; horses for about three, while standing up. Some animals can go very long periods without sleep. Some newborn dolphins and whales and their mothers stay awake for an entire month following birth.

All this variation makes it very hard to discover a single, universal function of sleep. Most sleep researchers now focus on the brain – things like memory consolidation and emotional processing – but sleep has important physical benefits, which is what we'll concentrate on here.

THE TRUTH ABOUT HOW MUCH SLEEP WE NEED

We all know eight hours is the magic number for a good night's sleep. However, nobody seems to know where that number came from. In questionnaires, people tend to say they sleep for between seven and nine hours a night, which might explain how eight hours has become a rule of thumb. But people also tend to overestimate how long they have been in the land of nod.

People living in traditional hunter-gather societies with no access to electricity generally sleep for six to seven hours a night and are perfectly fine on it.[1] So perhaps eight hours is the wrong target and we can get by just fine with seven. This seems to be a minimum requirement: a recent analysis in the US concluded that regularly getting less sleep than that increases the risk of obesity, heart disease, depression and early death, and recommended that all adults aim for at least seven hours.[2]

By this benchmark, recent reports suggest that we are walking around in a state of sleep deprivation. The US Centers for Disease Control and Prevention estimate that 35 per cent of US adults are getting less than seven hours a night, and a recent report by the UK's Royal Society for Public Health says Britons get an hour less than they need each night. A poll found that a third of adults experience symptoms of insomnia (for more on insomnia, see page 301).

The media often proclaim that we are getting less sleep than we used to, and there is a general perception that we are in the midst of an epidemic of sleep deprivation, and it is getting worse. Yet the numbers suggest this is an exaggeration. We may be collectively sleep deprived but there is no evidence that this is a recent phenomenon. A recent review of scientific records of sleep duration between 1960 and 2013 found no significant difference over time.[3]

The well-known 'fact' that people used to sleep around nine hours a night is a myth. The figure originates from a 1913 study of children, not adults. Even today, children continue to average this amount.

More support for the supposed epidemic of sleep deprivation comes from laboratory studies using very sensitive tests of sleepiness, such as the multiple sleep latency test, in which participants are sent to a quiet, dimly lit bedroom and instructed to 'relax, close your eyes and try to go to sleep'. These tests claim to reveal high levels of sleepiness in the population, but they are designed to squeeze out the very last drop of sleepiness, which, under everyday conditions, is largely unnoticeable.

Another line of evidence is that we typically sleep longer on holiday and at weekends, often up to nine or ten hours a night. It is often assumed that we do this to pay off a sleep debt built up during the week. However, just because we *can* sleep beyond our usual norm it doesn't necessarily follow that we need to. Why shouldn't we sleep purely for indulgence? We enthusiastically eat, drink and have sex beyond our biological needs. Why shouldn't it be the same with sleep?

In reality the amount of sleep we need varies from person to person and across lifespan. Newborns and infants sleep for up to seventeen hours a day. Some young adults need eleven hours while others only need six. It is often said that older adults need less sleep, but the evidence for that is weak.

Your genes influence the amount of sleep you personally need. Exactly which genes are involved is not well understood, but a recent study of over 50,000 people found one gene variant that added 3.1 minutes of sleep for every copy you have.[4]

Taking all of this into account, the US National Sleep Foundation updated their guidelines last year, and came up with a recommended range of seven to nine hours for adults, but with added leeway of an hour either side to account for natural variation.

So how much is enough for you? Having been you for the whole of your life, and with plenty of experience of sleeping – or not – you probably have a pretty good handle on how much sleep you ideally need, and also how much you can get by on.

But if you are still in doubt, a general rule is that you shouldn't need an alarm clock to wake you in the morning. If you're getting enough sleep you should wake naturally around the time you need to rise. If you regularly need an alarm to shock you into wakefulness you should consider going to sleep earlier. And that doesn't just mean going to bed earlier: time in bed doesn't equal time asleep.

Another test you can do is how long it takes you to fall asleep once you decide it is time. The ideal is ten to fifteen minutes, but if you are sleep deprived you'll drop off in about five. But measuring this is quite tricky as, by definition, the end point is loss of consciousness. Also be aware that you might not need as much sleep as you think. Many of us sleep for a long time out of habit, pleasure or boredom. So try cutting down and see how you feel.

While sleeping for the right amount of time is good for your health, take care not to overdo it – you can have too much of a good thing. The sweet spot for health seems to be about seven hours; regularly getting eight hours or more could send you to an early grave. Typically, the association between excess sleep and mortality is at least as strong as for sleep deprivation.

Why this is remains a mystery. It could be that people who are already ill sleep more, so it is not excess sleep per se that raises mortality but the underlying condition. Long sleep is associated with inflammation, an immune response linked to everything from depression to heart disease. Or it could be directly causal: when we are asleep we are moving very little, and there's plenty of evidence to show that inactivity is bad for you. Although this might not matter if you are active during the day, it could be that people who spend more time asleep do less exercise, possibly because they have less time.

Only a tiny minority of us, probably less than 3 per cent, can get by on four to six hours of sleep with no problems at all. This gift is almost certainly a result of genetics. Short sleep runs in families and researchers have identified a gene that, when engineered into mice, allows them to recover from sleep deprivation quicker and seems to enable them to whizz through the non-REM stages of sleep faster.[5]

But for the rest of us mere mortals, getting the right amount of sleep is vital for your physical well-being. Put simply, poor sleep can shorten your life.

THE TRUTH ABOUT GETTING TOO LITTLE SLEEP

Even though scientists argue about what sleep is for, they are pretty clear that getting the right amount is vital for your physical well-being. However, this is something that is often neglected.

On a routine visit to a doctor's surgery, you might expect to have your blood pressure and BMI measured, and be asked a few questions about diet, exercise and alcohol intake. One thing the doctor probably won't ask is how you sleep. If we're serious about preventive health, that is a serious oversight. Poor sleep is a major risk factor for obesity, diabetes, mood disorders and immune malfunction.

Sleep has been called the third pillar of good health, along with diet and exercise. But that's not doing it justice: sleep is the foundation on which these two other pillars rest. There is no physical or mental process that is not improved by sleep, or impaired by lack of it.

Lack of sleep has been linked with pretty much every major disease going, from heart disease to diabetes to cancer – even Alzheimer's, which we'll look at in more detail shortly. It interferes with your attention, working memory, planning, decision-making and time management. It reduces your ability to fight off infection. It makes you hungry and also depletes your willpower so you are more likely to overeat; chronic sleep deprivation has been linked to obesity. It affects your mood, leaving you feeling irritable and at greater risk of depression. And it causes accidents: driving on less than five hours' sleep trebles your risk of having a crash.

Being awake for twenty-four hours will leave you with the same level of cognitive impairment as having a blood alcohol content of 0.1 per cent – more than the drink-drive limit in several countries. But you don't have to be awake for that long to suffer the effects of sleeplessness. Just one night of poor sleep can have immediate effects. For example, a study limited healthy young adults to four hours of sleep for one night. The next day there was a measurable suppression of their immune systems.

Over the course of a few nights, chronic sleeplessness takes a greater toll on our bodies: in one study researchers allowed people just four hours of sleep a night for six nights in a row. They developed higher blood pressure, increased levels of the stress hormone cortisol and insulin resistance, a precursor to type 2 diabetes. They also produced half the normal number of antibodies in response to a flu vaccination.[6]

Fragmented sleep is also a problem, even if it adds up to a full eight hours. That's because when sleep is interrupted, you don't have time to go through all the restorative sleep stages.

Sleep is as vital for life as food or water. Laboratory rats deprived of sleep die within a month, and people can die of lack of sleep too, as shown by a rare genetic disease called fatal familial insomnia. The disease eventually robs victims of the ability to sleep and death follows within three months.

Unsurprisingly, given that we don't know the biological function of sleep, we still don't know why lack of sleep is deadly. However, this is not something you realistically need to worry about. The longer you stay awake the greater the pressure to sleep becomes. Without help you will be fighting hard to stay awake after thirty-six hours, and will find the urge to sleep overwhelming by forty-eight. Even then you've probably not been awake for two whole days. People who are severely sleep deprived often slip in and out of microsleeps – brief bouts of sleep lasting a second or two that happen without awareness, often with eyes wide open. That may be why being sleep deprived massively increases the risk of a car accident: drivers are briefly asleep at the wheel.

One of the most frightening diseases associated with lack of sleep is Alzheimer's, a form of dementia that currently affects more than forty-four million people worldwide and is predicted to grow as the population ages. Getting enough sleep is one of the most important factors determining whether you will develop the condition in the future.

As we age, sleep tends to get worse.[7] This is especially true for the quality of deep sleep. NREM is especially disrupted in people with Alzheimer's disease, and as the disease progresses, sleep disruption changes in parallel. We think this is because the probable causative agent of Alzheimer's is a toxic form of protein called beta-amyloid, which aggregates in sticky clumps (technically called plaques) in the brain. Amyloid plaques are poisonous to brain cells, impairing their function and killing them. Plaques preferentially build up in the middle part of the frontal lobe, a brain region essential for the generation of deep NREM sleep. The more amyloid in this area, the more impaired that person's deep sleep.

That is only half the story. Lack of sleep actually causes amyloid to build up in your brain, and so directly increases your risk of developing Alzheimer's disease.[8] This is because of the brain's waste disposal system, the glymphatic system. Just as the lymphatic system drains contaminants from your body, the glymphatic system uses cerebrospinal fluid to collect and break down harmful metabolic detritus generated by neurons as they go about their business.

Although the glymphatic system is active all the time, it kicks into high gear during deep NREM. At the same time the brain's glial cells also shrink 60 per cent, which creates greater space for the cerebrospinal fluid to clean out the gubbins. During deep sleep the glymphatic system expels up to twenty times more effluent than normal. In fact, this deep sleep clean may be one of the principal functions of sleep.[9]

One type of toxic debris washed away by the glymphatic system is amyloid protein, the toxin associated with Alzheimer's disease. In the laboratory, people prevented from going into deep sleep by sounds that rouse them slightly but don't wake them up, nudging them out of stage 3 sleep back into stage 2, show a build-up of amyloid in the spinal fluid. That happens even when their total sleep time is normal.

Inadequate sleep and Alzheimer's disease are therefore locked in a vicious circle. Without enough sleep, amyloid builds up in the brain, inhibiting deep NREM sleep and further preventing removal of amyloid. More amyloid, less deep sleep; less deep sleep, more amyloid, and so on.

This is why lack of sleep across your lifespan will significantly raise your risk of Alzheimer's disease. But the flip side is that by improving sleep, you should be able to reduce the risk. Clinical studies with middle-aged and older adults who have sleep disorders lend credence to this. Among those whose sleep problems were treated, the rate of cognitive decline slowed, delaying the onset of Alzheimer's by up to ten years.

Insufficient sleep is only one risk factor for Alzheimer's. Sleep alone will not defeat it. Nevertheless, prioritising sleep at any age is a clear way of lowering your risk of developing the disease. Which is as good a reason as any to start working on your sleep routine.

THE TRUTH ABOUT HOW TO SLEEP WELL

We may not know what sleep is for, but there's a huge body of knowledge about how to do it better. Sleep therapists talk about having good 'sleep hygiene', which means creating an environment that promotes sleep and minimises the risk of being rudely awakened.

There are a number of things you can do to improve your sleep hygiene.

Keep it regular

Try to stick to a strict sleep timetable, going to bed and waking up at the same time, even if you're (not) really tired, or don't have to get up the next morning. Wake-up time is the most important consideration as this will build up your sleep pressure during the day.

Keep it cool

Your body core temperature needs to drop by approximately 1.2 °C to get to sleep. This is why it's much harder to fall asleep in a room that's too hot than one that's too cold. About 18.5 °C is optimal, which is probably colder than you think. If your feet get cold, wear socks – it's not your peripheral but your core temperature that matters.

Studies show that people with sleep disorders who wake up a lot during the night can benefit from wearing a suit that slightly warms the skin. Counter-intuitively, this helps the body to release more heat. The cooling effect reduces the number of awakenings and also leads to more restorative slow-wave sleep. Taking a hot bath before bed can help achieve the same thing.

Keep it dim

Switch off as many lights as possible in the last hour before bed so as not to interfere with natural production of the sleep hormone melatonin, which is produced in the evening. Dim, red light before bed is best. Use table lamps rather than bright overhead lights. Install warm-toned light bulbs or invest in dimmable, colour-changing bulbs.

Tablets, phones and laptops generate lots of short wavelength blue light, which interferes with melatonin production. Using screens for two hours before bed reduces melatonin concentrations by 22 per cent. One experiment compared the sleep patterns of people who read a book on an iPad before bed with those who read it in print. After a few days, those reading on screen were taking longer to nod off and getting less REM sleep. Around 40 per cent of people in the US admit to taking their phone into bed and using it just before trying to get to sleep.

However, the dire warnings about screens being the mortal enemies of sleep are overblown. The effects are real, but subtle. Anything up to two hours won't do terrible damage to your melatonin levels. And if bedtime TV is your vice, relax. While the light from the box is bright, we normally watch from far enough away to avoid the melatonin effect.

During the night, keep it as dark as possible. People living in areas with high levels of light pollution tend to go to bed and wake up later than those in areas of mostly natural light. They also sleep less, are more tired during the day and are less satisfied with the quality of their sleep. That could be down to other factors, such as living in a busy and noisy city, but light seems a probable culprit. Investing in some blackout curtains can help aid a restful night.

Keep it quiet

Low-level sounds such as other people or pets moving about the house or traffic outside can interfere with sleep, whether you're aware of them or not. Even if they don't wake you up they may drag you out of deep sleep into a lighter phase. Invest in some comfy earplugs.

Keep it familiar

People often sleep badly when they are away from home. Brain scans of people sleeping in unfamiliar places found that parts of one hemisphere remain unusually active. This goes away once the sleeping place has become familiar. This 'first-night effect' may be an evolutionary adaptation, keeping you on alert to make sure the new environment is safe.

Keep trying

If you wake up in the night don't lie awake in bed for more than about twenty minutes. You need to reboot. Get up and do something quiet and relaxing such as a jigsaw or reading until the urge to sleep returns.

Keep it clean

Pass on the coffee, alcohol and cigarettes. Avoid caffeine after 1 p.m., and alcohol after 6 p.m. Smoking affects sleep no matter what time you light up. Each cigarette smoked during the day reduces total sleep time by 1.2 minutes. Animal studies suggest that nicotine disrupts a circadian clock protein in the lungs and brain.

Alcohol is a sedative, but sedation is not sleep. Unfortunately, it is easy to mistake one for the other, and if you regularly use alcohol to help you sleep you should think again, as it can be counterproductive. Having a few drinks before bed disrupts slow-wave sleep, adding a boost of alpha brainwaves that are usually only present in the daytime. Even a small early-evening tipple can be bad for sleep. The alcohol in a couple of drinks around happy hour will be metabolised by bedtime, but for some reason still causes a lot of extra wakefulness in the second part of the night. Older people are more sensitive to the effects.

Another danger of relying on alcohol as a sleep aid is that you then find you need a pick-me-up in the morning – often caffeine but also nicotine or something even stronger. Being over-caffeinated can lead to problems falling asleep, leading to alcohol use again. This vicious circle is called the sedative-stimulus loop, and is bad for both your sleep and your wakefulness.

If you fancy a drink in the evening try sour cherry juice. It's rich in melatonin, and a recent study found that after seven days, healthy adults who drank it twice a day got on average thirty-four minutes more sleep and napped less during the day. The US National Sleep Foundation (NSF) also recommends almonds, walnuts, bananas, pineapple, oranges and kiwi fruit as sources of melatonin.

Warm milk is a classic sleep promoter but the NSF says its effects are probably more psychological than physiological. If you believe something will promote sleep, then it probably will. The placebo effect is a powerful thing.

Foods rich in the amino acid tryptophan, meanwhile, increase serotonin levels, which sounds like it would be bad for sleep but may actually promote it. Eating tryptophan-rich foods has been shown to make people feel sleepier. The NSF says that low serotonin levels can contribute to insomnia and recommends eating cottage cheese before bed, perhaps sweetened with melatonin-rich fruits.

Cheese before bed – surely a recipe for nightmares, or at least disturbed sleep? Maybe not: in 2014 a trade association for UK cheese producers, called (amusingly) the British Cheese Board, tested the effect of evening cheese consumption on sleep and dreaming. It found that cheese eaters reported better-quality sleep than a placebo group. The volunteers also said they had fewer dreams than normal. But the study was not published in a journal, so should be taken with a pinch of cheese.

In any case high-fat foods – of which cheese is a fine example – have also been found to be associated with taking longer to fall asleep, perhaps because fat delays stomach emptying so slows down digestion, meaning your digestive system is active for longer in the night. Hence cottage cheese, the lowest-fat cheese of all. There are plenty of other low-fat, tryptophan-rich foods. Maybe try some lean chicken or prawns before bed.

Keep it natural
Prescription sleeping pills should be approached with caution. Paradoxically, taking them often makes sleep worse in the long run. They are also associated with higher rates of mortality and cancer.[10] Many are what sleep scientists call 'knock-out drops', which, like alcohol, promote sedation, not real sleep. If you think you need them, you must seek expert medical advice.

You can buy melatonin pills but they probably aren't the answer. Their half-life in the body is just thirty minutes to two hours, which might explain why studies into whether melatonin supplements can improve sleep in general produced mixed results.

There are plenty of herbal sleeping pills, but the evidence is mixed; ditto herbal teas such as chamomile and ginger. But there is reasonably good evidence from small studies that two herbs often used to promote sleep – valerian and kava – do decrease the time it takes to fall sleep and promote deeper sleep. Lavender, hops, lemon balm and passionflower are also touted as mild sleep aids but the evidence for these is weaker.

So there you have it. To optimise your sleep, don't stay up late, never have a lie-in, avoid caffeine and alcohol, drink sour juice, eat cottage cheese, make sure you're uncomfortably cold, turn off screens, drag yourself up in the middle of the night to do something boring and don't sleep anywhere unfamiliar such as a hotel or friend's house. No wonder most of us feel we're not getting enough sleep . . .

THE TRUTH ABOUT CATCHING UP ON MISSED SLEEP

Despite the benefits and undoubted pleasures of sleep, the demands and temptations of wakefulness often get in the way. So what happens when we burn the candle at both ends?

The acute effects of sleep deprivation are reversed when you catch up on the hours of sleep you have lost. If you've built up a sleep debt the solution is to repay it as soon as possible with a lie-in or a nap. This 'rebound sleep' is what many of us do with a weekend lie-in.

However, sleep deprivation can become a vicious circle. The more sleep deprived you become, the more you underestimate how tired you actually are. If you are suffering from chronic sleep loss – after a hectic period of work, for example – you probably need a holiday to break the cycle.

But there's a more serious concern. The jury is still out on whether repaying sleep debts can cancel out the long-term health effects of too little sleep. We know that shift work and jet lag, which mess with our body clock, also cause havoc with our health. Now, it seems, skimping on sleep in the week and catching up at the weekend, a phenomenon known as social jet lag, might cause the same kinds of health problems as shift work.

So, although anyone can recover from the short-term effects of the odd late night, a long-term habit of catching up on sleep at the weekends may well catch up with you in the end.

THE TRUTH ABOUT POWER NAPS

Caught napping? Clever you! Napping is no longer a sign of laziness, but a smart and efficient way to reap the rewards of sleep.

A ten-minute 'nano-nap' can boost alertness, concentration and attention for as much as four hours afterwards. Double it and you also increase your powers of memory and recall. Either way, you are unlikely to enter the deeper stages of sleep, so will avoid the phenomenon known as sleep inertia, the groggy feeling that can occur when awoken from deep sleep. On the flip side, you won't get the benefits of deep sleep, which provides the biggest boost to learning. If that's your aim, opt for a nap of between sixty and ninety minutes. Such a longer nap could also improve your positivity. A forty-five-minute nap should take you through a cycle of REM, and brain scans of people following a REM sleep nap showed more positive responses to images and to pleasant experiences.

However, light sleep turns out to be more important than we thought. During light sleep the brain emits short bursts of electrical activity called spindles that have been likened to popping champagne corks, which appear to have benefits for memory and learning.

There could be physical benefits too. Deep sleep lowers your blood pressure and lowers the contracting speed of the heart, both of which promote cardiovascular health. A large study of Greek men found that those who did not take a siesta had worse cardiovascular health and increased rates of cancer.[11]

If you're tempted, napping is easy. Find a warm, dim and quiet place where you can lie down (getting to sleep when you're sitting takes 50 per cent longer). If you want to keep it brief, drink a cup of coffee immediately beforehand – the caffeine kicks in after about twenty minutes, waking you up refreshed.

THE TRUTH ABOUT
SLEEP HACKING

For some people the goal is not to get more sleep, but less. These 'sleep hackers' get many extra hours of wakefulness by re-engineering their sleep patterns, foregoing full nights in favour of shorter sleeps and naps. This is essentially the fasting diet for sleep. Be careful, though. Just because dietary fasting works doesn't mean sleep fasting does too. The old adage that snoozers are losers turns out to be right, though for the wrong reasons.

One regime is called the Uberman sleep schedule. This allows you to sleep for no more than thirty minutes every four hours, giving you twenty-one hours of wakefulness every day and just three of sleep. It's one of a number of extreme sleep schedules that promise to maximise waking hours.

Another, the Everyman, allows one 'long' sleep of three hours plus three twenty-minute naps. The most extreme is Dymaxion, which rations daily sleep to just four thirty-minute naps. Can going without sleep to such an extreme be a good idea?

People who attempt the Uberman or other regimes often report a period of unpleasant adjustment, including headaches, tremors, anxiety and (surprise) tiredness. But after about a month some people adjust and feel normal again.

However, sleep researchers warn that hacking sleep is not a wise lifestyle choice. The body clock is hard-wired biologically, and for good reason. We know a full night's sleep allows our brain to cycle several times through a number of phases, each with restorative properties. Fragmented sleep has been shown to be just as damaging as total sleep deprivation for some health measures, such as metabolism.

People who regularly don't get enough sleep die younger. So for the extra hours you save, you might end up losing years from the end of your life.

According to some historians, however, breaking sleep into bouts is entirely natural. People in pre-industrial civilisations around the world naturally segmented their sleep into two distinct phases, with an hour or two of 'quiet wakefulness' in the middle of the night. This period was filled with bed-bound activities such as conversation and sex. We've done away with this practice, but by this logic, people who experience middle of the night insomnia may simply be expressing a natural sleep pattern.

But this claim remains controversial. If true you'd expect modern-day hunter-gatherers to follow this pattern, but none have yet been discovered to do so. Much like the rest of us, they generally prefer to stay up for at least three hours after sunset and then sleep in one big chunk. That doesn't mean that biphasic sleep does not happen, but it suggests it is not the natural pattern.

THE TRUTH ABOUT INSOMNIA

Anyone who has lain awake at night watching the minutes grind by and wondering why they can't sleep understands the misery of insomnia. Unfortunately, we really don't know much about what causes it or how to cure it.

An occasional inability to sleep is so common that it should perhaps be considered completely normal. One survey of adults in France found that 73 per cent had problems sleeping from time to time.[12]

Occasional sleep problems do not necessarily meet a clinical diagnosis of insomnia. To be a proper insomniac you have to be unable to fall or stay asleep for at least three nights a week, and your sleep disturbance has to have consequences for your waking life, such as excessive tiredness or impairment of normal functioning. The problem can be transient, lasting for just a few days; acute, over weeks; or chronic.

Even by these criteria, insomnia is very common. It affects one in five men and one in three women at some point in their adult lives, making it the world's most common sleep disorder.[13]

All that missed sleep has serious consequences. Insomnia has been linked to accidents and poor mental and physical health. It also causes economic damage because of reduced productivity and absenteeism.

Insomnia can be remedied if it is the result of a treatable condition such as restless legs syndrome or sleep apnoea. But for the majority of cases no underlying cause is found, and relief is hard to come by. The advice is usually to improve your sleep hygiene, by following some of the advice listed earlier in this chapter, such as reducing screen time or investing in blackout blinds. Drugs are another option, but they don't offer a long-term solution.

The key problem is that we still don't really understand insomnia. Is it psychological? Is it physical? Is it both? No one can agree. As for why women report poorer quality and more disrupted sleep, and have a 40 per cent higher risk of insomnia than men, again we don't know.

For some people, the cause of insomnia is worrying about insomnia. If you lie awake worrying that failure to sleep will have awful consequences in the morning, you are not going to get back to sleep. Follow the advice of the sleep hygiene experts: get up and do something to distract yourself from worry.

A lot of insomnia is actually all in the mind, a condition called 'sleep hypochondria'. About a quarter of people's perceptions of how well they sleep don't correlate with how they actually sleep. 'Complaining good sleepers' – people who believe they are insomniacs, even though their night-time brain activity suggests otherwise – are most likely to experience symptoms such as daytime fatigue, depression and anxiety. 'Non-complaining bad sleepers', by contrast, are remarkably free of ill effects.

If you're interested in measuring your own sleep duration and even how much of each variety you're getting, there are any number of apps and wearable technologies that can tell you. But beware: some researchers warn that these are leading to an epidemic of what has been dubbed 'orthosomnia', or the quest for perfect sleep. This is the sleep equivalent of orthorexia, a counterproductive obsession with eating a healthy diet.

Given that worry about poor sleep is a bigger problem than actual poor sleep, maybe you should just try to sleep without distraction.

But even that won't necessarily stop you from feeling tired.

THE TRUTH ABOUT FEELING TIRED ALL THE TIME

You've had another tiring day and, not for the first time this week, decide to turn in early. After a pretty good night's sleep you wake up naturally and feel . . . exhausted.

If this sounds depressingly familiar, you're not alone. According to a recent survey, about 30 per cent of visits to doctors involve complaints about being tired all the time.[14] Some 20 per cent of people in the US report having experienced fatigue intense enough to interfere with living a normal life.[15]

Being tired all the time (TATT) was once seen as simply a symptom of poor sleep or physical over-exertion, but no more. Sleep is a factor – the US Centers for Disease Control and Prevention estimate that 35 per cent of people are short on sleep – but isn't the only one. Fatigue can have many causes.

Chronic tiredness is a symptom of many common diseases, and also a hazard of ageing. But it can also descend on people without any apparent underlying cause.

A commonly cited reason is that life is more exhausting than ever. Caught between the demands of work and family, the ever-present pressure to exercise and the cornucopia of late-night home entertainment, it is perhaps no surprise so many of us feel run off our feet. Yet this is probably a fallacy.

People have always complained of being worn out, and harked back to the golden days when everyone slept for nine hours a night. In 1894, the *British Medical Journal* ran an editorial bemoaning the 'hurry and excitement, rush and worry, and anxiety and high tension' of modern life. But as we now know, the idea that nine hours was once the norm but has been rendered unattainable by modern life is an urban myth.

If modern life isn't to blame, another possibility is that at least some fatigue is down to a lack of sleep. However, researchers distinguish between sleepiness and fatigue, considering them to be related but subtly different. The good news is that there is an easy way to tell which applies to you. Called the sleep latency test, it is used widely in sleep clinics to assess whether people are sleep deprived. If you lie down somewhere quiet during the day and fall asleep within a few minutes, then you are either lacking sleep or potentially suffering from a sleep disorder. If you don't drop off within fifteen minutes or so, yet still feel tired, fatigue is the more likely problem.

So if it's not the same thing as sleepiness, what is fatigue? It seems that it is in part mental, not just physical. In experiments, people who were asked to lift weights while doing brain-taxing tasks such as mental arithmetic had 25 per cent less endurance than those who simply lifted weights. Subsequent imaging studies showed why: thinking hard lowers activity in frontal brain regions, which are involved in directing movements as well as having a hand in concentration. When the brain is stretched, it can make muscles tired too.

Another possibility is that daytime fatigue stems from a problem with the circadian clock, which regulates periods of mental alertness through the day and night. This regulation is the responsibility of the brain's suprachiasmatic nucleus (SCN), which coordinates hormones and brain activity to ensure that we generally feel alert by day and snoozy at night. Under normal circumstances, the SCN generates an alertness peak at the start of the day, a dip in the early afternoon and a shift to sleepiness in the evening.

The amount of sleep you get at night has little impact on this cycle. Instead, how alert you feel depends on the quality of the hormonal and electrical output signals from the SCN. The SCN sets its clock by the amount of light hitting the retina. Too little light in the mornings, or too much at night, can disrupt its signals, and either can lead to daytime fatigue. Experts worry that circadian rhythm disruption is already quite common, and getting more so with the increased use of light at night.

So if you spend your days feeling as if you have not yet woken up properly but are not sleepy at night, a badly calibrated SCN might be to blame. Try to spend at least twenty minutes outside every morning and limit evening screen time to avoid the SCN getting stuck in daytime mode.

The last thing fatigued people probably feel like doing is exercise but several studies have linked it to reduced fatigue, and it could be another way to reset your SCN.[16] This may explain why people who start exercising regularly often report sleeping better even though they don't actually sleep for any longer.[17] When it comes to sleep, quality may be more important than quantity.

Exercise also fights the flab, and there are good reasons to think that reducing fat levels could help tackle fatigue. Body fat not only takes more effort to carry around, but releases leptin, a hormone that signals to the brain that the body has adequate energy stores. Studies have linked higher leptin levels to greater perceived fatigue, a finding that makes perfect sense from an evolutionary perspective: if you have plenty of fat laid down, you don't need the motivation to go out and find food. Interestingly, people who fast regularly often report feeling more energetic than when they ate frequently.

There could be another reason why fat means fatigue. As we have seen, excess fat is linked to inflammation, a part of the immune response that rouses other parts into action by releasing cytokines into the bloodstream. Cytokines also make you feel drained, as anyone who has had a cold can attest. This feeling may be an evolved strategy to help fight off infections. When you need time to recuperate, fatigue is your friend.

Animal studies back this up. Mice given a drug that causes low-level inflammation avoid the running wheel, an activity they normally can't get enough of. If inflammation can rob mice of their vitality, there is good reason to suspect something similar holds for us.

Even if you're not fat or sick, inflammation could still be running you down. A sedentary lifestyle, chronic stress and poor diet have all been linked to low-level inflammation. If so, then some simple lifestyle changes could go a long way to perking you up: more exercise, some stress management and a better diet.

However, fatigue is often a vicious circle. Exercise and diet require motivation, but lack of motivation is a hallmark of fatigue. This, too, may be caused by inflammation. It lowers activity in two areas of the brain associated with motivation: the frontostriatal networks, which are involved in reward-based decision-making; and the insula, which processes bodily sensation of fatigue.

So fatigue is complicated. But that means there are a number of things you can try to stop being TATT. Some are standard-issue advice in health and fitness circles and keep on recurring throughout this book: diet, exercise and daylight.

Dehydration is also said to make you feel drained, and there is some evidence to support the idea. One study found that mild dehydration – a 1.5 per cent dip below the body's normal water volume, which can occur quite easily – can cause fatigue, particularly in younger women. It takes a 2 per cent drop in hydration to make us feel thirsty, so dehydration could be the cause of your fatigue. Making an effort to drink a bit of water even if you're not thirsty is worth a try.

Finally, there is one remedy for fatigue that you won't hear elsewhere, and which has the added benefit of being fun. If all else fails, good advice is to do something scary (if you like that sort of thing): the release of adrenaline could help you overcome lethargy. If you are stuck in the slow lane of endless fatigue, a bit of speed and excitement – perhaps a ride on a rollercoaster – could be just the tonic you need.

THE TRUTH ABOUT LIGHT AND SLEEP

One of the well-known elements of good sleep hygiene is to limit bright light in the evening and keep your room nice and dark. But it turns out that light is even more important than that, not just for good sleep but for good health in general.

Prior to the invention of gas lighting at the turn of the nineteenth century, the only artificial light our ancestors saw was from fire, candles or oil lamps. People spent many of their waking hours outside, and – moonlight notwithstanding – often passed the night in pitch darkness.

Today, we have a very different relationship with light. The average Westerner spends 90 per cent of their time indoors. When we do go outside, tall buildings often overshadow open spaces. That means we live in a kind of perpetual gloaming: not enough light during the day and too much at night. This unnatural exposure is increasingly being linked to poor sleep and disturbed circadian rhythms, the twenty-four-hour cycles in our biology and behaviour, with severe consequences for our health.

Too little sunlight is also contributing to a widespread deficiency in vitamin D that may be detrimental to our immune and cardiovascular systems (for more on the immune system, see page 247). The good news is that even small increases in your exposure to bright light during the day have a wide range of benefits.[18]

Light intensity is measured in lux, which refers to the amount of light striking a surface. On a cloudless summer day direct sunlight delivers something like 100,000 lux. Normal daylight gives us about 10,000 and an overcast day 1,000. Twilight comes in at 10 lux. Most indoor spaces are between overcast and twilight, with offices and classrooms clocking up around 200 lux. Even a metre from a window, light intensity dips below any form of daylight.

In other words, most of us spend our daytimes in the equivalent of near-twilight. On a summer's day, even if it is 50,000 lux outside, people in the UK are exposed to an average of 587 lux, and in winter that drops to 210.

Studies of the Amish, a religious community in Pennsylvania who avoid modern technology, show us what we're missing. In summer, the Amish are exposed to an average daytime illuminance of 4,000 lux. Even in winter the Amish get 1,500 lux. After sunset, the average illuminance in Amish homes is around 10 lux – up to five times lower than in electrified homes.

Light makes us feel good, and in recent years biologists have come to understand why. At the back of the eye, behind the image-forming rod and cone cells in the retina, are light-sensitive cells called intrinsically photosensitive retinal ganglion cells (ipRGCs). These fire in response to any bright light, and are particularly sensitive to blue light. This includes bright daylight, but also light from many LEDs and screens.

IpRGCs send signals to areas of the brain that control alertness and also a patch of brain tissue called the suprachiasmatic nucleus (SCN), which we've already met.[19] This is the body's master clock, which adjusts the auxiliary circadian clocks ticking away in every cell to keep them synchronised with one another and with external time. It interprets bright light as a signal that it is daytime. This system evolved for bright light exposure in the day, especially in the morning. It resets the body clock to 'daytime' mode and boosts alertness.

Exposure to daytime light also primes us for a good night. When Dutch researchers fitted twenty people with devices to record light exposure and then assessed their sleep, they found that higher daylight exposure was associated with more deep sleep and a lower incidence of fragmented sleep.[20] It seems that exposure to bright morning light can inoculate us against excessive light in the evening, especially the sort that beams out at us from our TV, computer and smartphone screens. A good dose of lux in the morning prevents your body clocks from shifting when exposed to blue light before bed.

Darkness at night is equally important. As we know, light at night suppresses the release of melatonin, a hormone that tells the body and brain that it is time to get ready for bed. Exposure to light in the evening also stimulates those ipRGCs at the back of the eye, boosting alertness just when you want to be winding down and fooling your master clock into thinking it is still daytime. Together, these conspire to make you less inclined to sleep. It's a bit like jet lag, where your body and brain are out of sync with the external world and you can't sleep even if you want to.

Experiencing a big contrast between day and night is also important. Light exposure influences not just the timing but also the amplitude of our circadian rhythms: under more constant light conditions the difference between day and night becomes flattened out, which is associated with poorer sleep.

Greater daylight exposure is also associated with lower scores on a scale of depression, which is consistent with other findings that bright morning light can help treat seasonal affective disorder and other forms of depression. That was assumed to be because light at the right time helps to improve quality of sleep. But the link may be more direct: a recent animal study showed that ipRGCs also connect to the thalamus, a brain area related to mood.[21]

Study after study has found positive effects of light exposure during the day. When researchers compared office workers' sleep, they found that staff who got more bright light during the day – either because they sat right next to a window, walked or cycled to work, or spent their breaks outdoors – fell asleep faster at night and slept for longer than those who got less light.[22] Workers exposed to more bright daylight between 8 a.m. and noon took an average of eighteen minutes to fall asleep at night, compared with forty-five minutes in the low-light exposure group. They slept for around twenty minutes longer and had fewer sleep disturbances.

In another very recent experiment twelve volunteers spent three nights living in glass domes in the Danish countryside, and were hence exposed to the twenty-four-hour light–dark cycle. In the morning the volunteers were significantly more alert and their drop-off in melatonin arrived an average twenty-six minutes earlier.

This all suggests some obvious changes we could make to our lifestyles, but unfortunately we are often not the masters of our own destiny. Many people spend all day inside buildings where, even though it is light enough to work, the illumination is nowhere near what they'd get outdoors.

So what can you do about it? Getting more sunlight in the morning is a good idea. Open your curtains when you wake up and eat breakfast somewhere bright. If you can walk or cycle to work, work next to a window or get outside during a break, then do so. If you have your own desk but are not right next to a window consider investing in a full-spectrum lamp. Swap indoor exercise for an outdoor equivalent.

You can also get light alarm clocks, which mimic the break of day and can help reset circadian rhythms. Light eases us out of deep sleep, leaving us less groggy even when modern life demands that we rise before dawn.

Evenings matter too. You may have heard that being exposed to blue light in the evening can throw a spanner into your circadian rhythms, stop you from sleeping and damage your health. This is based on solid science showing that intense, bluish-white light – aka daylight – suppresses the production of the sleep-promoting hormone melatonin. The same kind of light floods out from screens, leading to fears that watching TV and using phones and tablets in bed is doing horrible things to our circadian rhythms. Smartphones often come with a blue-light filter to allow you to watch the screen at night without worrying about screwing your melatonin levels.

Several studies have suggested that the light from smartphones can disrupt sleep, but if you think about it, this is quite implausible. Consider all those light-deprived office workers. If the light from screens were powerful enough to mimic daylight, they wouldn't be light deprived.

In fact, recent research shows that only larger devices produce enough bluish-white light to affect levels of melatonin.[23] The threshold for melatonin disruption is about an hour of 85 lux; a tablet iPad could provide that amount, but phones typically don't. However, the longer you stare the more likely you are to suppress melatonin.

For light sources emitting a warmer colour, such as incandescent light bulbs or warm-white LEDs, the levels found in a room in the evening are not enough to suppress melatonin production. Nor would a TV screen watched from two metres or more.

This emerging understanding of how light can profoundly affect our health, and what we can do about it, has led some researchers to predict that light will be the Next Big Thing in personal health, with wearable devices coming onto the market to track light exposure and nudge us to get more of it.

However, it is still unclear precisely how much daylight is necessary to optimise health, and it may well differ depending on what you are trying to achieve. Half an hour in the morning may be enough if you want to reset your circadian clock. If it's alertness you want, you may need bright light exposure for the whole day.

But as a general rule it is a bright idea to lighten up your days and darken your nights.

I'm not a good sleeper and have always had bouts of insomnia. I used to lie awake for hours, watching the minutes grind by while fretting about being too tired to function the next day. I once went to an important all-day meeting after a completely sleepless night spent in an increasingly desperate insomniacal tizzy. As it turned out I was fine.

I have now come to regard sleep as being a bit like water. If you need it, your body will tell you, and there's no point pushing yourself to get eight glasses or eight hours. Nowadays if I can't sleep I just shrug it off, think about whatever I'm working on at the time, and tell myself that sleep is naturally biphasic and that I'll fall back when my circadian clock is good and ready. More often than not, the next thing I know is that I've woken up and it's morning.

This isn't in the traditional sleep hygiene manual, but it works for me. And whatever I've previously said about that retort, I do think that when it comes to sleep, finding what works for you is the best way.

If you do struggle to sleep, it really is worth trying everything. The dangers of not getting enough are becoming increasingly clear, including the truly frightening prospect that it can lead to dementia. Sleep is rightly considered the foundation of good health. Almost everything in this book – from eating well to exercising – is easier if you've slept well.

CAN I LIVE
FOR EVER?

SPOILER ALERT: YOU are going to die. And, if you live long enough, you are going to get old first. That is not good news for your personal health and well-being. Not only is being old the main risk factor for death, it also raises the risk of numerous chronic diseases, including dementia, type 2 diabetes, cardiovascular disease, cancer and cataracts. These are known as the age-related diseases – around twenty are recognised – and they cluster together because they have the same underlying cause.

These start to hit you around the age of sixty, and their incidence increases exponentially from there onwards. Many elderly people have been diagnosed with two or more; simultaneously having five or six is not uncommon. These people are described as being in a state of late-life morbidity, from which there is usually only one way out.

The sobering truth is that once you are past forty-five or so, as far as your physical health is concerned it is downhill all the way to the end.

It does not have to be this way. Gerontologists talk about something called 'healthspan', which is perhaps even more desirable than lifespan. It refers to the ratio of healthy years (that is, discounting the years of being alive but in late-life morbidity) to overall lifespan. Think of it as 'quality-of-life span'.

How can you achieve good healthspan? It is time to reveal a secret. If you have got this far you have absorbed a lot of information about nutrition, diet, exercise, sleep, light, stress, pollution, supplements, medicines and more. It may all seem bewildering, but is all geared towards the same goal: to slow down the ageing process and increase healthspan. You may also increase your lifespan at the same time. Win-win.

Recall the Hadza people who we first met in the chapter about exercise and whose hunter-gatherer lifestyle delivers an ideal fitness regime? Not only are they physically fit and lean, they age extremely well. The diseases of later life so familiar to Westerners are almost unknown to the Hadza, and not because they die young. Late-life morbidity is the price we pay for our modern lifestyles.

So if you don't fancy spending your final years battling against an ever-growing burden of chronic diseases, then diet, exercise and all the rest are your friends.

THE SCIENCE OF AGEING . . .

The biology of ageing is essentially a progressive loss of the body's ability to repair itself, which is itself an unfortunate product of evolution. When we are young, damage depresses reproductive fitness, the ability to pass our genes on to our children and grandchildren, so has been selected against. But the rapid doubling of longevity has left evolution floundering to keep up. Keeping repair processes working for longer would probably increase our reproductive fitness – it would help us to raise more great- and great-great grandchildren, for example – but natural selection has not yet had time to get to work on the latter end of our lifespan.

As repair pathways fail, damage proliferates. Organs and tissues clog up with clumps of protein and other detritus. Mutations accumulate. Chromosomes start to unravel. Some cells become cancerous, others turn into zombies. Immune defences weaken.

Mitochondria, the powerhouses of our cells, fall into disrepair. Low-level inflammation – called 'inflammaging' – creeps throughout the body.

Soon enough this damage progresses to the classic age-related diseases: cardiovascular disease, atherosclerosis, osteoarthritis, cancer, dementia, type 2 diabetes, cataracts and more. And the older you get, the faster they hit. The incidence of almost every chronic disease suddenly increases exponentially around the age of sixty.

The loss of repair processes has other physical downsides that are less pathological but perhaps just as inconvenient and psychologically demoralising (depression is also recognised as a disease of old age). Muscle mass declines and fat accumulates on midriffs, backsides, thighs and around internal organs. Exercise becomes harder as heart and lung function decline. Joints stiffen and become painful. Injuries heal more slowly. Skin wrinkles and becomes more papery; wobbly bits sag. We sleep less well, see less well – most adults over forty-five have some degree of long-sightedness due to thickening and stiffening of the eye lens – and hear less well. Hair goes grey and falls out.

Thank science, then, that we now know enough about those repair processes to reboot them.

. . . AND HOW TO SLOW IT DOWN

For most of human history people have simply had to accept the inevitability of ageing. But that is changing. In recent years biologists have finally figured out what causes us to grow old and die. Not only do we increasingly understand that the classic lifestyle interventions such as diet and exercise are actually slowing down and even reversing ageing, there is a growing confidence that we can intervene even more directly.[1] Ageing is due to our biology, and this biology can be hacked.

Drug companies are busy translating the discoveries of ageing research into anti-ageing medicines and investors are betting big money on what many predict will become the biggest industry of all time. In as little as two years, the first scientifically validated anti-ageing drugs could be on the market.[2]

The interventions are not here yet, and will mostly take the form of pharmaceuticals and other prescribed medical interventions. But there are still lessons you can learn and even things you can do to reap the benefits.

If you feel like you have heard all this before, that is because you probably have. The fountain of youth legend has gripped humanity's imagination for centuries, and history is littered with the corpses of people who claimed ageing would be cured in their lifetimes or that immortality was within reach. Peddlers of snake oil and hype have promised much and delivered little. But this time it is different.

One key way it is different is that, in the past decade, biologists have proven that you can slow the ageing process using drugs – not just in experimental animals but in humans, too.[3] Several clinical trials are producing promising results against a range of age-related diseases such as osteoarthritis and immuno-senescence. These are not ageing per se but are caused by the underlying biology, so can be taken as proxies of ageing. There is good reason to believe that a drug that can alleviate osteo-arthritis will work for a broad range of age-related diseases.

Another important difference is that the goalposts have moved. Anti-ageing researchers used to seek extra years of life, but not any more. The aim now is to keep people healthier before they die. This is technically known as 'compression of morbidity' or 'extending healthspan' – the number of disease-free years. That means keeping life expectancy the same, eighty-five, but being healthy for eighty-four rather than decrepit for the final ten.

Extending healthspan would be an achievement on a par with the doubling in life expectancy that has happened over the past 150 years. That is obviously great, but healthspan has lagged behind. Between 2000 and 2015 global life expectancy rose by five years but the number of healthy years only rose by 4.6 years. An average of 20 per cent of life is now spent in late-life morbidity.

A lot of the smart money is on a class of drugs called senolytics, which search and destroy knackered cells that build up in tissues and organs as we age. These cells have suffered some sort of irreversible damage and entered a state called 'senescence' where they hit the emergency brake, hunker down and await destruction.

This process probably evolved to stop cells from becoming cancerous. But it eventually backfires. Senescent cells are normally cleared by white blood cells, but that process goes wrong during ageing. Uncleared cells are like zombies, beyond repair but undead and causing havoc. They pump out a range of inflammatory proteins, which are a major cause of inflammaging and damage the healthy tissues around them.

Senescent cells are now known to be a causative agent in many age-related diseases, including cancer, atherosclerosis, type 2 diabetes, osteoarthritis, Parkinson's, Alzheimer's and cataracts. Tellingly, if you take senescent cells from an old mouse and transplant them into a young mouse, it ages prematurely and acquires the diseases of old age. The opposite is also true. If you remove senescent cells from old mice they rejuvenate somewhat. And it appears to be never too late. Even very old, very decrepit mice benefit.

Senolytics are drugs that mimic this removal of senescent cells, by selectively targeting and killing the old cells.[4]

One great advantage of senolytic drugs is that people could take an occasional dose, perhaps once every six or twelve months, to clear out the trash. Most other anti-ageing drugs would probably have to be taken much more regularly, which raises obvious safety concerns about long-term side effects.

When might these drugs be available? The answer is: soon, maybe. The usual caveats about most experimental drugs falling by the wayside during clinical trials apply, but gerontologists say they could be in the clinic in as little as five years. You can even get them now, if you are prepared to take risks.

The drug of choice for biohackers is an occasional dose of the chemotherapy drug dasatinib plus plant extract quercetin, a combination that is being investigated in the laboratory. But be very careful – these drugs have not been tested in a human trial and biologists warn that they are not proven to be safe. Some senescent cells serve a useful purpose, in wound healing for example. A big challenge is to work out how to take out bad zombie cells without hitting the good ones. Do NOT take them unless you are prepared to risk your life.

There is a much safer way to destroy your zombies: exercise. It appears to exert its multiple health-giving effects in part because it is a senolytic.

Another area that may prove fruitful is a new take on an old anti-ageing strategy called caloric restriction, which we encountered in the chapter on diets and weight loss. Caloric restriction and periodic fasting have been shown to extend both lifespan and healthspan in every animal they have been foisted upon, probably because they activate some of the protective pathways that are progressively blunted by ageing.

One of those pathways is called 'MTOR', pronounced 'em-tor'.[5] It was discovered by scientists investigating the drug rapamycin, which was developed as an immunosuppressant for transplant patients but later turned out to extend lifespan in worms, flies and mice. They found that it exerts its effects by switching off a protein complex inside cells that they named 'mechanistic target of rapamycin' (MTOR).

MTOR evolved for surviving starvation. Its key function is nutrient sensing. If you eat, it is activated and tells the cell to grow and divide. If you don't eat, it switches off, which up-regulates protective pathways. These pathways include autophagy, the process by which cells scavenge dysfunctional organelles and molecules for energy – a waste disposal system for cells. MTOR thus slows down and reverses the accumulation of the gubbins that builds up in tissues, and hence retards the ageing process. Intriguingly, MTOR appears to become progressively stuck in the 'on' position as animals age.

Actual starvation is not necessary: rapamycin is just one of many drugs known to inhibit MTOR in mice even when started late in life. And just as with senolytics, the science is now being turned into therapies for humans. Most won't become available for years, and even then will be prescribed only to people diagnosed with specific age-related diseases. So if you were sceptical about trying intermittent fasting, then consider that it is a highly effective DIY for MTOR inhibition.

Some other popular anti-ageing biohacks may be targeting MTOR, though the actual mechanism is not clear. One of the most popular is the diabetes drug metformin, which has a good safety record and extends life and the duration of health in some animal models, although it recently failed a major test in mice. Other favourites are spermidine, a nutritional supplement that restores mitochondrial function in older mice, and nicotinamide mononucleotide, another compound shown to extend longevity in mice.

Again, approach these with caution: many are not proven human medicines. But spermidine is found in many foods, especially wheatgerm and the fermented Japanese soya-bean dish called natto.

Another option is to take one or other of the new generation of anti-ageing nutritional supplements that are coming onto the market. These are designed to switch off MTOR and have been shown to work in mice; they are also proven to be safe for human consumption. One is Rejuvant, made by the Florida-based company Ponce de Leon (named after a Spanish conquistador and explorer who supposedly sought the fountain of youth in Florida), and Basis, made by another US company, Elysium Health.

The final great hope of anti-ageing medicine is another modern take on an old discovery. Back in the 1970s, biologists began experimenting with a gruesome procedure called parabiosis, where the circulatory systems of two animals are plumbed together, sometimes for weeks on end, so they share blood. Parabiosis was developed to study conjoined twins but hinted at some intriguing effects on longevity. Old mice joined to young ones rejuvenate, while the young mice age prematurely.

More recent experiments have confirmed the effect.[6] Even just injecting young plasma – or plasma from human umbilical cords – has a rejuvenating effect on old mice.

This is the basis of the 'young blood' craze that has made headlines around the world, and which recently prompted the FDA to issue a strongly worded warning about unregulated plasma transfusion therapies. It said they have 'no proven clinical benefit . . . and there are risks associated with the use of any plasma product.' Drugs mimicking the effect are in the pipeline but as yet there is no safe or clinically proven way to benefit from parabiosis. Do not attempt it.

BIOLOGICAL AGE VERSUS CHRONOLOGICAL AGE

Everybody knows how old they are. But do you know how old you really are? You actually have two ages: your chronological age and your biological age. Think of it as the number of candles on your birthday cake versus the freshness of the cake itself.

Scientifically speaking, biological age is an attempt to measure how much an individual's cellular repair mechanisms have declined, which is now known to be the main cause of ageing. The number that comes out can diverge widely from your chronological age – in extreme cases by a decade or more – which is why some people seem younger than their years and others appear prematurely aged.

Differences between your chronological age and biological age are largely caused by lifestyle choices. The biggest factor is being fat; morbid obesity can add fifteen years. But most people are within five years of their biological age, either more or less.

Biological age is useful because it is often a more reliable indicator of future health than chronological age. For example, people with a higher biological age are at more risk of developing Alzheimer's and osteoporosis regardless of their chronological age. It can also be improved by lifestyle interventions – unlike chronological age, biological age can go down as well as up.

One crude way to estimate your biological age is to look in

the mirror. That is kind of a no brainer – we are all experts at judging somebody's age from their face. But beyond the obvious signifiers such as wrinkles and bags under the eyes there are some subtle markers of more rapid biological ageing. Older faces are characterised by wider mouths, increased distance between nose and mouth and, in men, a more protruding nose.

There are also more sophisticated techniques that use molecular markers to estimate biological age. The one considered the most accurate uses epigenetics, chemical modifications to DNA that occur during a person's lifetime and which switch genes on or off. It turns out that our cells contain an intrinsic epigenetic clock that is a good measure of biological age. Predictable changes in certain methylation patterns as people age can now be used to estimate it to within 2.9 years.

This clock is accelerated in the tissues of obese people, and ticks more slowly for those who regularly consume fish and vegetables and drink in moderation. Exercise also seems to toggle it.

Companies are now springing up that offer to measure your biological age from a saliva sample. Their services are not cheap and you need to ask about how accurate they are before shelling out. For example, one leading company in this area says it has an error rate of 4.75 years, which might lead you to question its usefulness. Say you are chronologically forty-five but your biological age comes back as forty-two. Great news – except that it might actually be as high as forty-seven, or as low as thirty-eight.

But if you want a baseline from which to launch your own personal rejuvenation campaign and also a benchmark to monitor your progress, a biological age test is worth thinking about.

Another way to find out your biological age is to subscribe to the services of a consumer epigenetics company, which routinely measure biological age alongside other health-related DNA markers (for more on epigenetic and genetic testing, see page 247).

HOW TO LIVE TO A RIPE OLD AGE

Healthspan and biological age are all well and good, but lifespan and chronological age matter too. After all, you've only got one life and if you want to make the most of it you need as much of it as you can get.

For many of us, getting past eighty is a very reasonable aspiration, but getting to a hundred is still very much a minority sport.

Perhaps you think you stand little chance of celebrating your hundredth birthday. Longevity depends in no small part on having the right genes, and few people win that particular lottery. However, bear in mind that centenarians are a fast-expanding demographic. Worldwide, there were an estimated 315,000 in 2012, but that had grown to 450,000 just three years later.[7]

This cannot be explained by genetics. It has to be environmental. In other words, there are lifestyle changes you can make that can expand your lifespan. Some are obvious and, at this point, hardly worth repeating. At the risk of labouring the point, everything in this book – on diet, exercise, hydration, sleep, prevention, stress and more – is ultimately aimed at increasing your healthspan and lifespan. But there is one more that is worth a mention before we call it a wrap.

Look on the bright side

There are surprisingly few things that long-lived individuals have in common, but one thing most share is an optimistic outlook on life. Researchers who work with centenarians often describe them as gregarious and fun to be with, and evidence is piling up that a sunny disposition is correlated with good health and a longer life.

For example, a study of nuns found that those who had the most positive outlook as young women were the healthiest in old age. Optimism improves the prospects of patients with heart conditions and infectious diseases. Older men with an optimistic outlook are half as likely to get cardiovascular disease as those with a negative one. It goes on.

It probably comes down to stress. Positive thinking lowers levels of the hormone cortisol, which dampens the immune response and also has adverse effects on the cardiovascular system and brain.

Some people are naturally laid-back, but even if you are more like a coiled spring there are things you can do to dial down your cortisol. These include meditation, yoga and t'ai chi. Even a deep breath can help. Exercise and good sleep will help too.

Another thing that centenarians have in common is that, at every stage of life, they are healthier than their peers in the same age cohort. In other words, being healthy now pays dividends in the future.

That is something to bear in mind when wrestling with your inbuilt gluttony and sloth. Whatever your starting point, getting a bit healthier today will mean that you are likely to live a bit longer tomorrow. You have to start somewhere. Make this the first day of your new life.

AFTERWORD

I STARTED THIS book with a confession, and I'll end it with another. I don't like self-help books. As a genre, I find them vapid and exploitative, playing on people's insecurities and promising quick, simplistic fixes for what are often complex problems.

But I have just written a self-help book, so am I not tarring myself with my own brush?

I hope you'll agree that this is a self-help book with a difference. It doesn't tell you that you are special or promise to unleash your inner potential. It doesn't promise to turn you into the rich, successful world-beater that you always knew you were. It doesn't offer inspirational quotes or ancient wisdom. It doesn't dish up a simple five-step plan for getting healthy; it doesn't offer a simple five-step plan for *anything*. In other words, it isn't your typical bundle of feel-good mumbo-jumbo.

This is a serious, evidence-based guide to living healthier and, we hope, longer. If you genuinely want to change your life for the better, it will work. Science is the most powerful tool we have for understanding the world we live in, including how to live a healthy and lengthy life. If something hasn't been tested scientifically and shown to work, then it is just another opinion. It might work. But it might not. If you want to take your chances, go ahead. But for every opinion there's some hard data. I know where I'm putting my time and money.

As always, however, there's a disclaimer. Scientific knowledge is, by definition, provisional. There's always a chance that it will be revised or superseded by a better theory or some new data. This is what makes science a uniquely powerful system of generating knowledge. Unlike, say, theology or philosophy, it does not claim to have all the answers but can accommodate new facts as they come in. This is how progress is made.

Consider the science we started with, the health effects of dietary fat. Twenty years ago most nutritionists would have put their money on 'case closed'. As new information came in, however, it became increasingly clear that there was something not quite right about the lipid hypothesis. But instead of sticking stubbornly to their guns, nutrition scientists assimilated the awkward data and revised their ideas accordingly. The result? A better lipid hypothesis; the old one wasn't wrong, by the way, just incomplete. The new one is probably incomplete too, and will, in time, be superseded by a better one. But for now it is the best thing we have.

Some people find the provisional nature of science an excuse to dismiss it all as 'just a theory', and pick and choose what they want to believe. This is the 'it worked for me' school of thought – something I often hear when I'm telling my friends and family about why their supplement/diet/exercise of choice, or the fad they've just read in the weekend paper, isn't supported by evidence. As is often said, anecdote is not the singular of data.

To be clear, I'm not entirely dismissing 'it worked for me'. Human bodies are complex and varied – but averaging out across large groups of people is the method of health science. So it is possible that something that doesn't work for the average human does work for you. It's possible, but unlikely. There's also the placebo effect, whereby merely believing that something will work can lead it actually to work. But, again, I'd always advocate following where the evidence leads.

If you want certainties and platitudes, the standard self-help book is for you. But if you want a nuanced scientific guide to healthy living, you're holding it.

Throughout the process of writing this book I have resisted calls to distil the evidence down to quick and simple take-homes. And with good reason – the scientists who work in these areas exhort us to consider the totality of the evidence, rather than cherry-pick.

However, there is one simple take-home that I can offer. If you see a health claim somewhere, ask yourself the following four questions: Does it sound too good to be true? Is the source credible? Is the evidence robust? Who stands to benefit?

I won't patronise you by telling you what answers you are looking for. If you've got this far, you're perfectly capable of working that out for yourself. Here's to your good health.

ACKNOWLEDGEMENTS

THIS BOOK WOULD not have been even remotely possible without my brilliant colleagues at *New Scientist*. Their skill, talent and dedication continue to amaze me even after working with them and their predecessors for nearly twenty years. Many of the chapters in this book are based on articles they either wrote or edited; for that reason I'm especially indebted to biomedical reporters and editors past and present: Andy Coghlan, Catherine de Lange, Kate Douglas, Alison George, Jessica Hamzelou, Rowan Hooper, Michael Le Page, Tiffany O'Callaghan, Penny Sarchet and Clare Wilson. Apologies if I have missed anybody, it has been a long haul. The subediting team also deserve a mention for their sterling work polishing up the original articles.

I'm also grateful to *New Scientist*'s senior leadership team, especially Nina Wright, Emily Wilson and Richard Webb, for their support, cajoling and patience.

The book also would not have happened without the persistence and vision of Georgina Laycock and Abigail Scruby at John Murray and *New Scientist*'s literary agent Toby Mundy.

Last, but not least, I must thank my wife Clare and sons Jonah and Jude, who yet again tolerated months of mental and physical absence and occasional grumpiness as the tough deadlines took their toll. Love you all xx

NOTES

The Truth About Food

1. Stephen, A. M. and Sieber, G. M. (1994). 'Trends in Individual Fat Consumption in the UK 1900–1985', *British Journal of Nutrition* 71, 5, 775–8. https://doi.org/10.1079/BJN19940183

2. Siri-Tarino, P. W, Sun, Q., Hu F. B. et al. (2010). 'Meta-analysis of Prospective Cohort Studies Evaluating the Association of Saturated Fat with Cardiovascular Disease', *American Journal of Clinical Nutrition*, 91,3, 535–46. https://doi.org/10.3945/ajcn.2009.27725

3. Chowdhury, R., Warnakula, S., Kunutsor, S. et al. (2014). 'Association of Dietary, Circulating, and Supplement Fatty Acids With Coronary Risk: a Systematic Review and Meta-analysis', *Annals of Internal Medicine* 160, 6, 398–406. https://doi.org/10.7326/M13–1788

4. de Oliveira, O., Mozaffarian D., Kromhout D. et al. (2012). 'Dietary Intake of Saturated Fat by Food Source and Incident Cardiovascular Disease: the Multi-Ethnic Study of Atherosclerosis', *American Journal of Clinical Nutrition*, 96, 2, 397–404. https://doi.org/10.3945/ajcn.112.037770

5. Gardner, C. D., Kiazand, A., Alhassan, S. et al. (2007). 'Comparison of the Atkins, Zone, Ornish, and LEARN Diets for Change in Weight and Related Risk Factors Among Overweight Premenopausal Women. The A TO Z Weight Loss Study: A Randomized Trial', *Journal of the American Medical Association* 297,9,969–77.https://doi.org/I10.1001/jama.297.9.969

6. Shai, I., Schwarzfuchs, D., Henkin, Y. et al (2008). 'Weight Loss with a Low-Carbohydrate, Mediterranean, or Low-Fat Diet', *New England Journal of Medicine*, 359, 3, 229–41. https://doi.org/10.1056/NEJMoa0708681

7. Eddy, D., Schlessinger, L., Kahn, R. et al. (2009). 'Relationship of Insulin Resistance and Related Metabolic Variables to Coronary Artery Disease: A Mathematical Analysis', *Diabetes Care* 32, 2, 361–66. https://doi.org/10.2337/dc08-0854

8. See, for example, Lenoir, M., Serre, F., Cantin, L. and Ahmed, S. H. (2007). 'Intense Sweetness Surpasses Cocaine Reward', *PLoS ONE* 2, 8, e698. https://doi.org/10.1371/journal.pone.0000698

9. See, for example, Te Morenga, L., Mallard, S. and Mann, J. (2013). 'Dietary Sugars and Body Weight: Systematic Review and Meta-analyses of

Randomised Controlled Trials and Cohort Studies', *BMJ* 346, e7492. https://doi.org/10.1136/bmj.e7492

10. See, for example, Strazzullo, P., D'Elia, L., Kandala, N-B. and Cappuccio, F. P. (2009). 'Salt Intake, Stroke, and Cardiovascular Disease: Meta-analysis of Prospective Studies', *BMJ* 339, b4567. https://doi.org/10.1136/bmj.b4567

11. Cook, N. R., Cutler, J. A., Obarzanek, E. et al. (2007). 'Long-term Effects of Dietary Sodium Reduction on Cardiovascular Disease Outcomes: Observational Follow-up of the Trials of Hypertension Prevention' (TOHP), *BMJ* 334, 885. https://doi.org/10.1136/bmj.39147.604896.55

12. Kieneker, L. M., Eisenga, M. F., Gansevoort M.T. et al. (2018). 'Association of Low Urinary Sodium Excretion With Increased Risk of Stroke', *Mayo Clinic Proceedings* 93, 12, 1803–09. https://doi.org/10.1016/j.mayocp.2018.05.028

13. World Cancer Research Fund/American Institute for Cancer Research (2007). *Food, Nutrition, Physical Activity, and the Prevention of Cancer: A Global Perspective.* Washington, D.C. http://www.dietandcancerreport.org

14. 'Red Meat and Processed Meat' (2015), *IARC Monographs on the Evaluation of Carcinogenic Risks to Humans* 114.

15. Lippia, G., Mattiuzzi, C. and Cervellin, G. (2016). 'Meat Consumption and Cancer Risk: a Critical Review of Published Meta-analyses', *Critical Reviews in Oncology/Hematology* 97, 1, 1–14. https://doi.org/10.1016/j.critrevonc.2015.11.008

16. Pan, A., Sun, Q., Bernstein, A. M. et al. (2012). 'Red Meat Consumption and Mortality: Results From 2 Prospective Cohort Studies', *Annals of Internal Medicine* 172, 7, 555–63. https://doi.org/10.1001/archinternmed.2011.2287

17. Kappeler R., Eichholzer, M and Rohrmann, S. (2013). 'Meat Consumption and Diet Quality and Mortality in NHANES III', *European Journal of Clinical Nutrition* 67, 6, 598–606. https://doi.org/10.1038/ejcn.2013.59

18. Rohrmann, R., Overvad, K., Bueno-de-Mesquita, H. B, et al. (2013). 'Meat Consumption and Mortality: Results from the European Prospective Investigation into Cancer and Nutrition', *BMC Medicine* 11, 63. https://doi.org/10.1186/1741-7015-11-63

19. Ibid.

20. Michaëlsson, K., Wolk, A., Langenskiöld, S. et al. (2014). 'Milk Intake and Risk of Mortality and Fractures in Women and Men: Cohort Studies', *BMJ* 349, g6015. https://doi.org/10.1136/bmj.g6015

21. Cui, X., Zuo, P., Zhang, Q. et al (2006). 'Chronic Systemic D-galactose Exposure Induces Memory Loss, Neurodegeneration, and Oxidative

Damage in Mice: Protective Effects of R-Alpha-lipoic Acid', *Journal of Neuroscience Research* 83, 8, 1584–90. https://doi.org/10.1002/jnr.20899

22. Lampe J. W. (2011). 'Dairy Products and Cancer', *Journal of the American College of Nutrition* 30, 5, Suppl. 1, 464S–70S. https://doi.org/10.1080/07315724.2011.10719991

23. Elwood, P. C., Givens, D. I, Beswick, A. D. et al. (2008). 'The Survival Advantage of Milk and Dairy Consumption: an Overview of Evidence from Cohort Studies of Vascular Diseases, Diabetes and Cancer', *Journal of the American College of Nutrition*, 27, 6, 723S–734S. https://doi.org/10.1080/07315724.2008.10719750

24. Aune, D., Giovannucci, E., Boffetta, P. et al. (2017). 'Fruit and Vegetable Intake and the Risk of Cardiovascular Disease, Total Cancer and All-cause Mortality – a Systematic Review and Dose-response Meta-analysis of Prospective Studies', *International Journal of Epidemiology* 46, 3, 1029–56. https://doi.org/10.1093/ije/dyw319

25. Verhoeven, D. T, Goldbohm, R. A., van Poppel, G. et al. (1996). 'Epidemiological Studies on Brassica Vegetables and Cancer Risk', *Cancer Epidemiology, Biomarkers and Prevention* 5, 9, 733–48. https://cebp.aacrjournals.org/content/5/9/733.short

26. See, for example, De Carvalho, F. G., Ovídio, P. P., Padovan, G. J. et al. (2014). 'Metabolic Parameters of Postmenopausal Women after Quinoa or Corn Flakes Intake – a Prospective and Double-blind Study', *International Journal of Food Science and Nutrition* 65, 3, 380–5. https://doi.org/10.3109/09637486.2013.866637

27. Cassidy, A., Mukamal, K. J., Liu, L. et al. (2013). 'High Anthocyanin Intake is Associated with a Reduced Risk of Myocardial Infarction in Young and Middle-Aged Women', *Circulation* 127, 188–96. https://doi.org/10.1161/CIRCULATIONAHA.112.122408

28. Scheers, N., Rossander-Hulthen, L., Torsdottir, I. et al. (2016). 'Increased Iron Bioavailability from Lactic-fermented Vegetables is Likely an Effect of Promoting the Formation of Ferric Iron $(Fe3+)$', *European Journal of Nutrition* 55, 1, 373–82. https://doi.org/10.1007/s00394-015-0857-6g

29. See, for example, Hamet, M. F., Medrano, M., Pérez, P. F. et al. (2016). 'Oral Administration of Kefiran Exerts a Bifidogenic Effect on BALB/c Mice Intestinal Microbiota', *Beneficial Microbes* 7, 2, 237–46. https://doi.org/10.3920/BM2015.0103

30. Jun, L., Qi, L. and Yu, C (2015). 'Consumption of Spicy Foods and Total and Cause Specific Mortality: Population Based Cohort Study', *BMJ* 351, h3942. https://doi.org/10.1136/bmj.h3942

31. Heskey, C., Oda, K. and Sabaté, J. (2019). 'Avocado Intake, and Longitudinal

Weight and Body Mass Index Changes in an Adult Cohort', *Nutrients* 11, 3, 691. https://doi.org/10.3390/nu11030691

32. Hemler, E. C., Chavarro, J. E. and Hu, F. B. (2018). 'Organic Foods for Cancer Prevention – Worth the Investment?', *JAMA Internal Medicine* 178, 12, 1606–07. https://doi.org/10.1001/jamainternmed.2018.4363

33. González, N., Marquès, M., Nadal, M. and Domingo, J. L. (2019). 'Occurrence of Environmental Pollutants in Foodstuffs: A Review of Organic Vs. Conventional Food', *Food and Chemical Toxicology*, 125, 370–75. https://doi.org/10.1016/j.fct.2019.01.021

34. Hurtado-Barroso, S., Tresserra-Rimbau, A., Vallverdú-Queralt, A. and Lamuela-Raventós, R.M. (2017). 'Organic Food and the Impact on Human Health', *Critical Reviews in Food Science and Nutrition* 59, 4, 704–14. https://doi.org/10.1080/10408398.2017.1394815

35. Stergiadis, S., Leifert, C., Seal, C. J. et al. (2012). 'Effect of Feeding Intensity and Milking System on Nutritionally Relevant Milk Components in Dairy Farming Systems in the North East of England', *Journal of Agricultural and Food Chemistry*, 60, 29, 7270–81. https://doi.org/10.1021/jf301053b

36. Estruch, R., Ros, E., Salas-Salvadó, J. et al. (2013). 'Primary Prevention of Cardiovascular Disease with a Mediterranean Diet', *New England Journal of Medicine* 368, 1279–90. https://doi.org/10.1056/NEJMoa1200303

37. Estruch, R., Ros, E. and Salas-Salvadó, J. (2018). 'Primary Prevention of Cardiovascular Disease with a Mediterranean Diet Supplemented with Extra-Virgin Olive Oil or Nuts', *New England Journal of Medicine*, 378, e34. https://doi.org/10.1056/NEJMoa1800389

38. Bonaccio, M., Di Castelnuovo, A., Pounis, G. et al. (2017). 'High Adherence to the Mediterranean Diet is Associated with Cardiovascular Protection in Higher but not in Lower Socioeconomic Groups: Prospective Findings from the Moli-sani Study', *International Journal of Epidemiology* 46, 5, 1478–87. https://doi.org/10.1093/ije/dyx145

39. Menotti, A., Puddu, P. E., Lanti, M. et al. (2014). 'Lifestyle Habits and Mortality from all and Specific Causes of Death: 40-year Follow-up in the Italian Rural Areas of the Seven Countries Study', *Journal of Nutrition, Health and Aging* 18, 314–21. https://doi.org/10.1007/s12603-013-0392-1

40. Scientific Advisory Committee on Nutrition \(2019). Saturated fats and health. https://www.gov.uk/government/publications/saturated-fats-and-health-sacn-report

The Truth About Diets and Weight Loss

1. Betts, J. A., Richardson, J. D., Chowdhury, E. A. et al. (2014). 'The Causal Role of Breakfast in Energy Balance and Health: a Randomized Controlled

Trial in Lean Adults', *American Journal of Clinical Nutrition* 100, 2, 539–47. https://doi.org/10.3945/ajcn.114.083402

2. Betts, J. A., Chowdhury, E. A. and Gonzalez, J. T. (2016). 'Is Breakfast the Most Important Meal of the Day?', *Proceedings of the Nutrition Society* 75, 4, 464–74. https://doi.org/10.1017/S0029665116000318

3. Conley, M., Le Fevre, L., Haywood, C. et al. (2018). 'Is Two Days of Intermittent Energy Restriction Per Week a Feasible Weight Loss Approach in Obese Males? A Randomised Pilot Study', *Nutrition and Dietetics* 75,1, 65–72. https://doi.org/10.1111/1747-0080.12372

4. Wei, M., Brandhorst, S., Shelehchi, M. et al. (2017). 'Fasting-mimicking Diet and Markers/Risk Factors for Aging, Diabetes, Cancer, and Cardiovascular Disease', *Science Translational Medicine* 9, 377, eaai8700. https://doi.org/10.1126/scitranslmed.aai8700

5. Chaix, A., Zarrinpar, A., Miu, P. and Panda, S. (2014). 'Time-Restricted Feeding is a Preventative and Therapeutic Intervention against Diverse Nutritional Challenges', *Cell Metabolism* 20, 6, 991–1005. https://doi.org/10.1016/j.cmet.2014.11.001

6. Fontana, L., Cummings, N. E. and Arriola Apelo, S. I. (2016). 'Decreased Consumption of Branched-Chain Amino Acids Improves Metabolic Health', *Cell Reports* 16, 520–30. https://doi.org/10.1016/j.celrep.2016.05.092

7. Klein, A. V. and Kiat, H. (2015). 'Detox Diets for Toxin Elimination and Weight Management: a Critical Review of the Evidence', *Journal of Human Nutrition and Dietetics* 28, 6, 675–86. https://doi.org/10.1111/jhn.12286

8. Hue, O., Marcotte, J., Berrigan, F. et al. (2006). 'Increased Plasma Levels of Toxic Pollutants Accompanying Weight Loss Induced by Hypocaloric Diet or by Bariatric Surgery', *Obesity Surgery* 16, 9, 1145–54. https://doi.org/10.1381/096089206778392356

9. Eaton S. B. and Konner, M. (1985). 'Paleolithic Nutrition: a Consideration of its Nature and Current Implications', *New England Journal of Medicine* 312, 5, 283–9. https://doi.org/10.1056/NEJM198501313120505

10. Thompson, R. C., Allam, A. H., Lombardi, G. P. et al. (2013). 'Athero-sclerosis Across 4000 Years of Human History: the Horus Study of Four Ancient Populations', *Lancet* 381, 9873, 1211–22. https://doi.org/10.1016/S0140-6736(13)60598-X

11. Winston J., Craig, W. J. (2009). 'Health Effects of Vegan Diets', *American Journal of Clinical Nutrition* 89, 5, 1627S–1633S. https://doi.org/10.3945/ajcn.2009.26736N

12. Skodje, G. I., Sarna, V. K., Minelle, I. H. et al. (2018). 'Fructan, Rather Than Gluten, Induces Symptoms in Patients With Self-Reported

Non-Celiac Gluten Sensitivity', *Gastroenterology* 154, 3, 529–39.e2. https://doi.org/10.1053/j.gastro.2017.10.040

13. Francis, C. Y. and Whorwell, P. J. (1994). 'Bran and Irritable Bowel Syndrome: Time For Reappraisal,' *Lancet* 344, 8914, 39–40.

14. Jensen, M. D., Ryan, D. H., Apovian, C. M. and Ard, J. D. (2014). '2013 AHA/ACC/TOS Guideline for the Management of Overweight and Obesity in Adults: a Report of the American College of Cardiology/American Heart Association Task Force on Practice Guidelines and The Obesity Society', *Circulation* 129, S102–S138. https://doi.org/10.1161/01.cir.0000437739.71477.ee

15. Yannakoulia, M., Poulimeneas, D., Mamalaki, E. and Anastasiou, C. A. (2019). 'Dietary Modifications for Weight Loss and Weight Loss Maintenance', *Metabolism* 92, 153–62. https://doi.org/10.1016/j.metabol.2019.01.001

16. Flegal, K. M., Kit, B. K., Orpana, H. et al. (2013). 'Association of All-Cause Mortality With Overweight and Obesity Using Standard Body Mass Index Categories: a Systematic Review and Meta-analysis', *JAMA* 309, 1, 71–82. https://doi.org/10.1001/jama.2012.113905

17. Murakami, K., Sasaki, S., Takahashi, Y. et al. (2007). 'Hardness (Difficulty of Chewing) of the Habitual Diet in Relation to Body Mass Index and Waist Circumference in Free-living Japanese Women Aged 18–22 y.' *American Journal of Clinical Nutrition* 86, 1, 206–13. https://doi.org/10.1093/ajcn/86.1.206

18. Evenepoel, P., Geypens, B., Luypaerts, A. et al. (1998). 'Digestibility of Cooked and Raw Egg Protein in Humans as Assessed by Stable Isotope Techniques', *Journal of Nutrition* 128, 10, 1716–22. https://doi.org/10.1093/jn/128.10.1716

19. See, for example, Schmidt, S. L., Harmon, K. A., Sharp, T. A. et al. (2012). 'The Effects of Overfeeding on Spontaneous Physical Activity in Obesity Prone and Obesity Resistant Humans', *Obesity*, 20, 11, 2186–93. https://doi.org/10.1038/oby.2012.103

The Truth About Vitamins and Supplements

1. Bjelakovic, G., Nikolova, D., Gluud, L. L. et al. (2012). 'Antioxidant Supplements for Prevention of Mortality in Healthy Participants and Patients with Various Diseases', *Cochrane Database of Systematic Reviews* 3. https://doi.org/10.1002/14651858.CD007176.pub2

2. Martí-Carvajal, A. J., Solà, I., Lathyris, D. et al. (2013). 'Homocysteine-lowering Interventions for Preventing Cardiovascular Events', *Cochrane*

Database of Systematic Reviews 1. https://doi.org/10.1002/14651858.CD006612.pub3

3. Hemilä, H. and Chalker, E. (2013). 'Vitamin C for Preventing and Treating the Common Cold', *Cochrane Database of Systematic Reviews* 1. https://doi.org/10.1002/14651858.CD000980.pub4

4. Bolland, M. J., Barber, P. A., Doughty, R. N. et al. (2008). 'Vascular Events in Healthy Older Women Receiving Calcium Supplementation: Randomised Controlled Trial', *BMJ* 336, 262–66. https://doi.org/10.1136/bmj.39440.525752.BE

5. Bolland, M. J., Avenell, A., Baron, J. A. et al. (2010). 'Effect of Calcium Supplements on Risk of Myocardial Infarction And Cardiovascular Events: Meta-Analysis', *BMJ* 341, c3691. https://doi.org/10.1136/bmj.c3691

6. Tian, H., Guo, X., Wang, X. et al. (2013). 'Chromium picolinate supplementation for overweight or obese adults', *Cochrane Database of Systematic Reviews* 11, CD010063. https://doi.org/10.1002/14651858.CD010063.pub2

7. European Food Safety Authority Panel on Dietetic Products, Nutrition and Allergies (NDA). (2010). 'Scientific Opinion on the substantiation of health claims related to coenzyme Q10', *EFSA Journal* 10, 8, 1793. https://doi.org/10.2903/j.efsa.2010.1793

8. Schöttker, B., Jorde, R., Peasey, A. et al. (2014). 'Vitamin D and mortality: meta-analysis of individual participant data from a large consortium of cohort studies from Europe and the United States', *BMJ* 348, g3656. https://doi.org/10.1136/bmj.g3656

9. Abdelhamid, A. S., Brown, T. J., Brainard, J. S. et al. (2018). 'Omega-3 fatty acids for the primary and secondary prevention of cardiovascular disease', *Cochrane Database of Systematic Reviews* 7, CD003177. https://doi.org/10.1002/14651858.CD003177.pub3

10. Miller, E. R. 3rd, Juraschek, S., Pastor-Barriuso, R. et al. (2010). 'Meta-analysis of folic acid supplementation trials on risk of cardiovascular disease and risk interaction with baseline homocysteine levels.' *American Journal of Cardiology* 106, 4, 517–27. https://doi.org/10.1016/j.amjcard.2010.03.064

11. Towheed, T., Maxwell, L., Anastassiades, T.P. et al. (2005). 'Glucosamine therapy for treating osteoarthritis', *Cochrane Database of Systematic Reviews* 2, CD002946. https://doi.org/10.1002/14651858.CD002946.pub2

12. Shaw, K. A., Turner, J. and Del Mar, C. (2002). 'Tryptophan and 5-Hydroxytryptophan for depression', *Cochrane Database of Systematic Reviews* 1, CD003198. https://doi.org/10.1002/14651858.CD003198

13. European Food Safety Authority Panel on Dietetic Products, Nutrition

and Allergies (NDA). (2010). 'Scientific Opinion on the substantiation of health claims related to methylsulphonylmethane (MSM)'. *EFSA Journal* 8, 10, 1746. https://doi.org/10.2903/j.efsa.2010.1746

14. Rees, K., Hartley, L., Day, C. et al. (2013). 'Selenium supplementation for the primary prevention of cardiovascular disease,' *Cochrane Database of Systematic Reviews* 1, CD009671. https://doi.org/10.1002/14651858. CD009671.pub2

15. Singh, M. and Das, R. R. 'Zinc for the common cold', *Cochrane Database of Systematic Reviews* 6, CD001364 https://doi.org/10.1002/14651858. CD001364.pub4

16. Bjelakovic, G., Nikolova, D., Gluud, L. L. et al. (2007). 'Mortality in Randomized Trials of Antioxidant Supplements for Primary and Secondary Prevention: Systematic Review and Meta-analysis', *JAMA* 297, 8, 842–57. https://doi.org/10.1001/jama.297.8.842

17. Klein, E. A., Thompson, I. M., Tangen, C. M. et al. (2011). 'Vitamin E and the Risk of Prostate Cancer: The Selenium and Vitamin E Cancer Prevention Trial (SELECT),' *JAMA* 306, 14, 1549–56. https://doi.org/10.1001/jama.2011.1437

The Truth About Drink (and Drugs)

1. Valtin, H. (2002). '"Drink at Least Eight Glasses of Water a Day." Really? Is There Scientific Evidence for "8 × 8"?', *American Journal of Physiology-Regulatory, Integrative and Comparative Physiology* 283, 5, R993–R1004. https://doi.org/10.1152/ajpregu.00365.2002

2. Furst, H., Hallows, K. R., Post, J. et al. (2000). 'The Urine/Plasma Electrolyte Ratio: A Predictive Guide to Water Restriction', *The American Journal of the Medical Sciences* 319, 4, 240–244. https://doi.org/10.1016/S0002-9629(15)40736-0

3. Popkin, B. M., D'Anci, K. E. and Rosenberg, I. H. (2010). 'Water, Hydration, and Health', *Nutrition Reviews* 68, 8, 439–458. https://doi.org/10.1111/j.1753-4887.2010.00304.x

4. Bernabéa, E., Vehkalahti, M. M., Sheiham, A. et al. (2014). 'Sugar-Sweetened Beverages and Dental Caries in Adults: A 4-Year Prospective Study', *Journal of Dentistry* 42, 8, 952–958 https://doi.org/10.1016/j.jdent.2014.04.011

5. O'Connor, L., Imamura, F., Lentjes, M.A.H. et al. (2015). 'Prospective associations and population impact of sweet beverage intake and type 2 diabetes, and effects of substitutions with alternative beverages', *Diabetologia* 58, 1474. https://doi.org/10.1007/s00125-015-3572

6. Heneghan, C., Howick, J., O'Neill, B. et al. (2012). 'The evidence

underpinning sports performance products: a systematic assessment', *BMJ Open* 2, e001702. https://doi.org/10.1136/bmjopen-2012-001702

7. Pritchett, K. and Pritchett, R. (2013). 'Chocolate Milk: A Post-Exercise Recovery Beverage for Endurance Sports', *Acute Topics in Sport Nutrition* 59, 127–134. https://doi.org/10.1159/000341954

8. Rogers, P. J., Hogenkamp, P. S., de Graaf, C. et al. (2016). 'Does Low-Energy Sweetener Consumption Affect Energy Intake And Body Weight? A Systematic Review, Including Meta-Analyses, of the Evidence From Human and Animal Studies', *International Journal of Obesity* 40, 381–94. https://doi.org/10.1159/000341954

9. Loftfield, E., Cornelis, M. C., Caporaso, N. et al (2018). 'Association of Coffee Drinking With Mortality by Genetic Variation in Caffeine Metabolism: Findings From the UK Biobank', *JAMA Internal Medicine* 178, 8, 1086–97. https://doi.org/10.1001/jamainternmed.2018.2425

10. Boehm, K., Borrelli, F., Ernst, E. et al. (2009). 'Green Tea (*Camellia sinensis*) for the Prevention of Cancer', *Cochrane Database of Systematic Reviews* 3, CD005004. https://doi.org/10.1002/14651858.CD005004.pub2

11. Neturi, R. S., Srinivas, R., Simha, V. et al. (2014). 'Effects of Green Tea on Streptococcus mutans Counts – A Randomised Control Trail', *Journal of Clinical and Diagnostic Research* 11, ZC128 – ZC130. https://doi.org/10.7860/JCDR/2014/10963.5211

12. Coles, L. T. and Clifton, P. M. (2012). 'Effect Of Beetroot Juice On Lowering Blood Pressure In Free-Living, Disease-Free Adults: A Randomized, Placebo-Controlled Trial', *Nutrition Journal* 11, 106. https://doi.org/10.1186/1475-2891-11-106

13. Kohn, J. B. (2015). 'Is Vinegar an Effective Treatment for Glycemic Control or Weight Loss?', *Journal of the Academy of Nutrition and Dietetics* 115, 7, 1188. https://doi.org/10.1016/j.jand.2015.05.010

14. GBD 2016 Alcohol Collaborators (2016). 'Alcohol Use and Burden for 195 Countries and Territories, 1990–2016: A Systematic Analysis for the Global Burden of Disease Study 2016', *Lancet* 392, 10152, 1015–35. https://doi.org/10.1016/S0140-6736(18)31310-2

15. Ibid.

16. Wood, A. M., Kaptoge, S., Butterworth, A. S. et al. (2018). 'Risk Thresholds for Alcohol Consumption: Combined Analysis of Individual-Participant Data for 599,912 Current Drinkers in 83 Prospective Studies', *Lancet* 391, 10129, 1513–23. https://doi.org/10.1016/S0140-6736(18)30134-X

17. Mehta, G., Macdonald, S., Cronberg, A. et al. (2018). 'Short-Term Abstinence From Alcohol and Changes in Cardiovascular Risk Factors, Liver Function Tests and Cancer-Related Growth Factors: A Prospective

Observational Study', *BMJ Open* 8, e020673. https://doi.org/10.1136/bmjopen-2017-020673

18. de Visser, R. O., Robinson, E. and Bond, R. (2016). 'Voluntary Temporary Abstinence From Alcohol During "Dry January" and Subsequent Alcohol Use', *Health Psychology* 35, 3, 281–9. https://doi.org/10.1037/hea0000297

19. Perreault, K., Bauman, A., Johnson N. et al. (2017). 'Does Physical Activity Moderate the Association Between Alcohol Drinking and All-Cause, Cancer and Cardiovascular Diseases Mortality? A Pooled Analysis of Eight British Population Cohorts', *British Journal of Sports Medicine* 51, 651–7. https://doi.org/10.1136/bjsports-2016-096194

20. Penning, R., McKinney, A. and Verster, J. C. (2012). 'Alcohol Hangover Symptoms and Their Contribution to the Overall Hangover Severity', *Alcohol and Alcoholism* 47, 3, 248–52, https://doi.org/10.1093/alcalc/ags029

21. Köchling, J., Geis, B., Wirth, S. et al. (2019). 'Grape or Grain but Never the Twain? A Randomized Controlled Multiarm Matched-Triplet Crossover Trial of Beer and Wine', *The American Journal of Clinical Nutrition* 109, 2, 345–352. https://doi.org/10.1093/ajcn/nqy309

22. Pirie, K., Peto, R., Reeves, G. K. et al. (2013). 'The 21st Century Hazards of Smoking and Benefits of Stopping: A Prospective Study of One Million Women in the UK', *Lancet* 381, 9861, 133 –141. https://doi.org/10.1016/S0140-6736(12)61720-6

23. Nutt, D. J., King, L. A. and Phillips, L. D. on behalf of the Independent Scientific Committee on Drugs (2010). 'Drug Harms in the UK: A Multicriteria Decision Analysis', *Lancet* 376, 9752, 1558–1565. https://doi.org/10.1016/S0140-6736(10)61462-6

24. O'Brien, M. S. and Anthony, J. C. (2005). 'Risk of Becoming Cocaine Dependent: Epidemiological Estimates for the United States, 2000–2001', *Neuropsychopharmacology* 30, 1006–18. https://doi.org/10.1038/sj.npp.1300681

The Truth About Exercise

1. Morris, J. N. and Crawford, M. D. (1958). 'Coronary Heart Disease and Physical Activity of Work', *BMJ* 2, 5111, 1485–96. https://dio.org/10.1136/bmj.2.5111.1485

2. See, for example, Pedersen, B. K. (2019). 'The Physiology of Optimizing Health with a Focus on Exercise as Medicine', *Annual Review of Physiology* 81, 607–27. https://doi.org/10.1146/annurev-physiol-020518-114339

3. Diabetes Prevention Program Research Group (2002). 'Reduction in the

Incidence of Type 2 Diabetes with Lifestyle Intervention or Metformin', *New England Journal of Medicine* 346, 393–403. https://doi.org/10.1056/NEJMoa012512

4. Dunstan, D. W., Howard, B., Healy, G. N. et al. (2012). 'Too Much Sitting – Health Hazard', *Diabetes Research and Clinical Practice* 97, 3, 368–76. https://doi.org/10.1016/j.diabres.2012.05.020

5. Lear, S., Hu, W., Rangarajan, S. et al. (2017). 'The Effect of Physical Activity on Mortality and Cardiovascular Disease in 130,000 People from 17 High-Income, Middle-Income, and Low-Income Countries: The PURE Study', *Lancet* 390, 10113, 2643–2654. https://doi.org/10.1016/S0140-6736(17)31634-3

6. O'Donovan, G., Lee, I., Hamer, M. et al. (2017). 'Association of "Weekend Warrior" and Other Leisure Time Physical Activity Patterns With Risks for All-Cause, Cardiovascular Disease, and Cancer Mortality', *JAMA Internal Medicine* 177, 3, 335–42. https://doi.org/10.1001/jamainternmed.2016.8014

7. Bouchard, C, An, P., Rice, T. et al. (1999). 'Familial Aggregation of VO2 Max Response to Exercise Training: Results from the HERITAGE Family Study', *Journal of Applied Physiology* 87, 3, 1003–08. https://doi.org/10.1152/jappl.1999.87.3.1003

8. Tigbe, W. W., Granat, M. H., Satta, N. and Lean, M. E. J. (2017). 'Time Spent in Sedentary Posture is Associated with Waist Circumference and Cardiovascular Risk', *International Journal of Obesity* 41, 689–96. https://www.nature.com/articles/ijo201730

9. See, for example, Rees-Punia E., Evans, E. M., Schmidt, M. D. et al. (2019). 'Mortality Risk Reductions for Replacing Sedentary Time With Physical Activities', *American Journal of Preventative Medicine* 56, 5, 73–41. https://doi.org/10.1016/j.amepre.2018.12.006

10. Tabata, I., Nishimura, K., Kouzaki, M. et al. (1996). 'Effects of Moderate-intensity Endurance and High-intensity Intermittent Training on Anaerobic Capacity and VO2max', *Medical Science Sports Exercise* 28, 10, 1327–30. https://www.ncbi.nlm.nih.gov/pubmed/8897392

11. Ruiz, J. R., Sui, X., Lobelo, F. et al. (2008). 'Association Between Muscular Strength and Mortality in Men: Prospective Cohort Study', *BMJ* 337, a439. https://doi.org/10.1136/bmj.a439

12. See, for example, Swift, D. L., Johannsen, N. M., Laviec C. J. et al. (2014). 'The Role of Exercise and Physical Activity in Weight Loss and Maintenance', *Progress in Cardiovascular Diseases* 56, 4, 441–7. https://doi.org/10.1016/j.pcad.2013.09.012

13. Willbond, S. M., Laviolette, M. A., Duval, K., Doucet, E. (2010). 'Normal

Weight Men and Women Overestimate Exercise Energy Expenditure', *Journal of Sports Medicine and Physical Fitness* 50, 4, 377–84.

14. Dugas, L. R., Kliethermes, S., Plange-Rhule, J. et al. (2017). 'Accelerometer-measured Physical Activity is Not Associated with Two-year Weight Change in African-origin Adults from Five Diverse Populations', *PeerJ* 5, e2902. https://doi.org/10.7717/peerj.2902

15. Pontzer, H., Durazo-Arvizu, R., Dugas, L. et al. (2015). 'Constrained Total Energy Expenditure and Metabolic Adaptation to Physical Activity in Adult Humans', *Current Biology* 26, 3, 410–17. https://doi.org/10.1016/j.cub.2015.12.046

16. Paravidino, V. B., Mediano, M. F. F., Hoffman D. J. and Sichieri, R. (2016). 'Effect of Exercise Intensity on Spontaneous Physical Activity Energy Expenditure in Overweight Boys: A Crossover Study', *PLoS ONE* 11, 1, e0147141. https://doi.org/10.1371/journal.pone.0147141

17. Levine, J. A., Lanningham-Foster, L. M and McCrady, S. K. (2005). 'Interindividual Variation in Posture Allocation: Possible Role in Human Obesity', Science 307, 584–6 https://doi.org/10.1126/science.1106561

18. Moffitt, T. E., Arseneault, L., Belsky. D. et al. (2011). 'A Gradient of Childhood Self-control Predicts Health, Wealth, and Public Safety', *PNAS* 108, 7, 2693–8. https://doi.org/10.1073/pnas.1010076108

19. Baumeister, R. F., Bratslavsky, E., Muraven, M. and Tice, D. M. (1998). 'Ego Depletion: is the Active Self a Limited Resource?', *Journal of Personality and Social Psychology* 74, 5, 1252–65. https://doi.org/10.1037/0022-3514.74.5.1252

20. Job, V., Dweck, C. S. and Walton, G. M. 'Ego Depletion – Is It All in Your Head?: Implicit Theories About Willpower Affect Self-Regulation', *Psychological Science* 21, 11, 1686–93.

21. Miller, E. M., Walton, G. M., Dweck, C. S., Job, V. et al. (2012). 'Theories of Willpower Affect Sustained Learning', *PLoS ONE* 7, 6, e38680. https://doi.org/10.1371/journal.pone.0038680

22. Neal, D. T., Wood, W. and Drolet, A. (2013). 'How do People Adhere to Goals When Willpower is Low? The Profits (and Pitfalls) of Strong Habits', *Journal of Personal and Social Psychology* 104, 6, 959–75. https://doi.org/10.1037/a0032626

23. Crum, A. J. and Langer, E. J. 'Mind-Set Matters: Exercise and the Placebo Effect', *Psychological Science* 18, 2, 165–71.
https://doi.org/10.1111/j.1467-9280.2007.01867.x

24. Zahrt, O. H. and Crum, A. J. (2017). 'Perceived Physical Activity and Mortality: Evidence from Three Nationally Representative U.S. Samples', *Health Psychology* 36, 11, 1017–25. https://doi.org/10.1037/hea0000531

25. Crum, A. J., Corbin, W., Brownell, R. et al. (2011). 'Mind Over Milkshakes: Mindsets, Not Just Nutrients, Determine Ghrelin Response', *Health Psychology* 30, 4, 424–9.

26. Cassady, B. A., Considine, R. V. and Mattes, R. D. (2012). 'Beverage Consumption, Appetite, and Energy Intake: What Did You Expect?', *American Journal of Clinical Nutrition* 95, 3, 587–93, https://doi.org/10.3945/ajcn.111.025437

The Truth About Staying Well

1. https://digital.nhs.uk/data-and-information/publications/statistical/health-survey-for-england/health-survey-for-england-2013

2. Ridker, P. M., Danielson, E., Fonseca, F. A. H. et al. (2008). 'Rosuvastatin to Prevent Vascular Events in Men and Women with Elevated C-Reactive Protein', *New England Journal of Medicine* 359, 2195–207. https://doi.org/10.1056/NEJMoa0807646

3. Collins, R., Reith, C., Emberson, J. et al. (2016). 'Interpretation of the Evidence for the Efficacy and Safety of Statin Therapy', *Lancet* 388, 2532–61. https://doi.org/10.1016/S0140-6736(16)31357-5

4. Spence, J. D. and Dresser, G. K. (2016). 'Overcoming Challenges With Statin Therapy', *Journal of American Heart Association* 5, 1, e002497. https://doi.org/10.1161/JAHA.115.002497

5. Hira, R. S., Kennedy, K., Nambi, V. et al. (2015). 'Frequency and Practice-Level Variation in Inappropriate Aspirin Use for the Primary Prevention of Cardiovascular Disease: Insights From the National Cardiovascular Disease Registry's Practice Innovation and Clinical Excellence Registry', *Journal of the American College of Cardiology* 65, 2, 111–21. https://doi.org/10.1016/j.jacc.2014.10.035

6. J. Cuzick, M. A., Thorat, C., Bosetti, P. H. et al. (2015). 'Estimates of Benefits and Harms of Prophylactic Use of Aspirin in the General Population', *Annals of Oncology* 26, 1, 47–57. https://doi.org/10.1093/annonc/mdu225

7. Hannaford, P. C. and Iversen, L. (2013). 'Mortality Among Oral Contraceptive Users: an Evolving Story', *European Journal of Contraception & Reproductive Health Care* 18, 1, 1–4. https://doi.org/10.3109/13625187.2012.723226

8. Martin, R. M., Donovan, J. L. and Turner, E. L. (2018). 'Effect of a Low-Intensity PSA-Based Screening Intervention on Prostate Cancer Mortality: the CAP Randomized Clinical Trial', *JAMA* 319, 9, 883–95. https://doi.org/10.1001/jama.2018.0154

9. Tawakol, A., Ishai, A. and Takx, R. A. P. (2007). 'Relation Between Resting

Amygdalar Activity and Cardiovascular Events: a Longitudinal and Cohort Study', *Lancet* 389, 10071, 834–45. https://doi.org/10.1016/S0140-6736(16)31714-7

10. Müller, G., Harhoff, R., Rahe, C. et al. (2018). 'Inner-city Green Space and its Association with Body Mass Index and Prevalent Type 2 Diabetes: a Cross-sectional Study in an Urban German City', *BMJ*, 8, e019062. https://doi.org/10.1136/bmjopen-2017-019062

11. https://www.gov.uk/government/collections/monitor-of-engagement-with-the-natural-environment-survey-purpose-and-results

12. Nuckton, T. J., Koehler, E. A. and Schatz, S. P. (2014). 'Characteristics of San Francisco Bay Cold-Water Swimmers', *Open Sports Medicine Journal*, 8, 1–10. https://doi.org/10.2174/1874387001408010001]

13. https://www.rcplondon.ac.uk/projects/outputs/every-breath-we-take-lifelong-impact-air-pollution

14. Maher, B. A., Ahmed, I. A. M., Karloukovski, V. et al (2016). 'Magnetite Pollution Nanoparticles in the Human Brain', *PNAS* 113, 39, 10797–801. https://doi.org/10.1073/pnas.1605941113

15. Lewis, O. J., Killin, J. M., Starr, I. J. et al.(2016). 'Environmental Risk Factors for Dementia: a Systematic Review', *BMC Geriatrics* 16, 175. https://doi.org/10.1186/s12877-016-0342-y

16. Holt-Lunstad, J., Smith, T. B., Layton, J. B. (2010). 'Social Relationships and Mortality Risk: A Meta-analytic Review', *PLoS Med* 7, 7, e1000316. https://doi.org/10.1371/journal.pmed.1000316

17. Richardson, T., Elliott, P. and Roberts, R. (2017), 'Relationship Between Loneliness and Mental Health in Students', *Journal of Public Mental Health*, 16, 2, 48–54. https://doi.org/10.1108/JPMH-03-2016-0013

18. Cole, S. W., Hawkley, L. C., Arevalo, J. M. et al. (2007). 'Social Regulation of Gene Expression in Human Leukocytes', *Genome Biology* 8, R189. https://doi.org/10.1186/gb-2007-8-9-r189

19. Christensen, K. (2019). 'Longevity Enriched Families. How Did They Succeed?' Conference presentation, 3rd Interventions in Aging, Nassau, Bahamas.

20. Russell, D., Peplau, L. A. and Ferguson, M. L. (1978). Developing a measure of loneliness. *Journal of Personality Assessment* 42, 3, 290–294. https://doi.org/10.1207/s15327752jpa4203_11

21. Lindqvist, P. G., Epstein, E., Nielsen, K. et al. (2016). 'Avoidance of Sun Exposure as a Risk Factor for Major Causes of Death: a Competing Risk Analysis of the Melanoma in Southern Sweden Cohort', *Journal of Internal Medicine* 280, 4, 375–87. https://doi.org/10.1111/joim.12496

22. Liu, D., Fernandez, B. Hamilton, A. et al. (2014). 'UVA Irradiation of

Human Skin Vasodilates Arterial Vasculature and Lowers Blood Pressure Independently of Nitric Oxide Synthase', *Journal of Investigative Dermatology* 134, 7, 1839–46. https://doi.org/10.1038/jid.2014.27

23. Dennis, L. K., Vanbeek, M. J., Freeman, L. E. et al. (2008). 'Sunburns and Risk of Cutaneous Melanoma: Does Age Matter? A Comprehensive Meta-analysis', *Annals of Epidemiology* 18, 8, 614–27. https://doi.org/10.1016/j.annepidem.2008.04.006.

24. Autier, P., Mullie, P., Macacu, A. et al. (2017). 'Effect of Vitamin D Supplementation on Non-skeletal Disorders: a Systematic Review of Meta-analyses and Randomised Trials', *Lancet Diabetes Endocrinology* 5, 12, 986–1004. https://doi.org/10.1016/S2213-8587(17)30357-1

The Truth About Sleep

1. Yetish, G., Kaplan, H., Gurven, M. et al. (2015). 'Natural Sleep and Its Seasonal Variations in Three Pre-industrial Societies', *Current Biology* 25, 21, 2862–8. https://doi.org/10.1016/j.cub.2015.09.046

2. Consensus Conference Panel. (2015). 'Recommended Amount of Sleep for a Healthy Adult: a Joint Consensus Statement of the American Academy of Sleep Medicine and Sleep Research Society', *Sleep* 38, 6, 843–4. https://doi.org/10.5665/sleep.4716

3. Bin, Y. S., Marshall, N. S. and Glozier, N. (2012). 'Secular Trends in Adult Sleep Duration: a Systematic Review', *Sleep Medical Review* 16, 3, 223–30. https://doi.org/10.1016/j.smrv.2011.07.003

4. Gottlieb, D. J., Hek, K. and Chen T. H. (2015). 'Novel Loci Associated with Usual Sleep Duration: the CHARGE Consortium Genome-Wide Association Study', *Molecular Psychiatry* 20, 10, 1232–9. https://doi.org/10.1038/mp.2014.133

5. He, Y., Jones, C. R., Fujiki, N. et al. (2009). 'The Transcriptional Repressor DEC2 Regulates Sleep Length in Mammals', *Science* 325, 5942, 866–70. https://doi.org/10.1126/science.1174443

6. For a review of the evidence, see Bryant, P. A., Trinder, J. and Curtis, N. (2004). 'Sick and Tired: Does Sleep have a Vital Role in the Immune System?', *Nature Reviews Immunology* 4, 457–67. https://doi.org/10.1038/nri1369

7. Mander, B. A, Winer, J. R. and Walker, M. P. (2017). 'Sleep and Human Aging', *Neuron* 94, 1, 19–36. https://doi.org/10.1016/j.neuron.2017.02.004

8. Mander, B. A., Marks, S. M., Vogel, J. W. et al. (2015). 'β-amyloid Disrupts Human NREM Slow Waves and Related Hippocampus-dependent Memory Consolidation', *Natural Neuroscience* 18, 7, 1051–7. https://doi.org/10.1038/nn.4035

9. Xie, L., Kang, H., Xu, Q. et al. (2013). 'Sleep Drives Metabolite Clearance from the Adult Brain', *Science* 342, 6156, 373–7. https://doi.org/10.1126/science.1241224

10. Kripke, D. F., Langer, R. D. and Kline, L. E. (2012). 'Hypnotics' Association with Mortality or Cancer: a Matched Cohort Study', *BMJ* 2, e000850. https://doi.org/10.1136/bmjopen-2012-000850

11. Naska, A., Oikonomou, E., Trichopoulou, A. et al. (2007). 'Siesta in Healthy Adults and Coronary Mortality in the General Population', *Archives of Internal Medicine* 167, 3, 296–301. https://doi.org/10.1001/archinte.167.3.296

12. Leger, D., Guilleminault, C., Dreyfus, J. P. et al. (2009). 'Prevalence of Insomnia in a Survey of 12,778 Adults in France', *Journal of Sleep Research* 9, 35–42.

13. 'Extent and Health Consequences of Chronic Sleep Loss and Sleep Disorders' in *Sleep Disorders and Sleep Deprivation: An Unmet Public Health Problem*, ed. H. R. Colten and M. Altevogt, National Academies Press, 2006. https://www.nap.edu/read/11617/chapter/5

14. van't Leven, M., Zielhuis, G. A., van der Meer, J. W. et al. (2010). 'Fatigue and Chronic Fatigue Syndrome-like Complaints in the General Population', *European Journal of Public Health* 20, 3, 251–7. https://doi.org/10.1093/eurpub/ckp113

15. Kroenke, K. and Price, R. K. (1993). 'Symptoms in the Community: Prevalence, Classification, and Psychiatric Comorbidity', *Archives of Internal Medicine* 153, 21, 2474–80. https://doi.org/10.1001/archinte.1993.00410210102011

16. Puetz T. W., Flowers, S. S. and O'Connor, P. J. (2008). 'A Randomized Controlled Trial of the Effect of Aerobic Exercise Training on Feelings of Energy and Fatigue in Sedentary Young Adults with Persistent Fatigue', *Psychotherapy and Psychosomatics* 77, 167–174. https://doi.org/10.1159/000116610

17. Wong, S. N., Halaki, M. and Chow, C-M. (2013). 'The Effects of Moderate to Vigorous Aerobic Exercise on the Sleep Need of Sedentary Young Adults', *Journal of Sports Sciences* 31, 4, 381–6. https://doi.org/10.1080/02640414.2012.733823

18. Münch, M., Nowozin, C., Regente J. et al. (2013). 'Blue-Enriched Morning Light as a Countermeasure to Light at the Wrong Time: Effects on Cognition, Sleepiness, Sleep, and Circadian Phase', *Neuropsychobiology* 74, 201–18. https://doi.org/10.1159/000477093

19. Beaven, C. M. and Ekström, J. (2013). 'A Comparison of Blue Light and Caffeine Effects on Cognitive Function and Alertness in Humans', *PLoS One* 8, 10, e76707. https://doi.org/10.1371/journal.pone.0076707

20. Gordijn, M., Geerdink, M., Wams, E. et al. (2018). 'Relationship Between Daytime Light Exposure and EEG Sleep Architecture, Slow-wave Activity, and Sleep Quality in Young Healthy Office Workers', *Journal of Sleep Research* 27, 09.2018.
    http://hdl.handle.net/11370/e648236e-e31d-4e45-9626-9f7ee9c72bfb

21. Fernandez, D. C., Fogerson, M., Ospri, L. L. et al. (2018). 'Light Affects Mood and Learning through Distinct Retina-Brain Pathways', *Cell* 175, 1, 71–84, E18. https://doi.org/10.1016/j.cell.2018.08.004

22. Figueiro, M. G., Steverson, B., Heerwagen, J. et al (2017). 'The Impact of Daytime Light Exposures on Sleep and Mood in Office Workers', *Sleep Health* 3, 3, 204–15. https://doi.org/10.1016/j.sleh.2017.03.005

23. Nagare, R., Rea, M. S., Plitnick, B. and Figueiro, M. G. (2019). 'Nocturnal Melatonin Suppression by Adolescents and Adults for Different Levels, Spectra, and Durations of Light Exposure', *Journal of Biological Rhythms* 34, 2, 178–94. https://doi.org/10.1177/0748730419828056

Can I Live For Ever?

1. Partridge, L., Deelen, J. and Slagboom, E. (2018). 'Facing up to the Global Challenges of Ageing', *Nature* 561, 45–56. https://doi.org/10.1038/s41586-018-0457-8

2. Mullard, A. (2018). Anti-ageing Pipeline Starts to Mature', *Nature Reviews Drug Discovery* 17, 609–12. https://doi.org/10.1038/nrd.2018.134

3. Mahmoudi, S., Xu, L. and Brunet, A. (2019). 'Turning Back Time with Emerging Rejuvenation Strategies', *Nature Cell Biology* 21, 32–43. https://doi.org/10.1038/s41556-018-0206-0

4. See, for example, Xu, M., Pirtskhalava, T., Farr, J. N. et al. (2018). 'Senolytics Improve Physical Function and Increase Lifespan in Old Age', *Nature Medicine* 24, 1246–56. https://doi.org/10.1038/s41591-018-0092-9

5. Kennedy, B. K. and Lamming, D. W. (2016). 'The Mechanistic Target of Rapamycin: the Grand ConducTOR of Metabolism and Aging', *Cell Metabolism* 23, 6, 990–1003. https://doi.org/10.1016/j.cmet.2016.05.009

6. See, for example, Loffredo, F. S., Steinhauser, M. L., Jay, S. M. et al. (2013). 'Growth Differentiation Factor 11 Is a Circulating Factor that Reverses Age-Related Cardiac Hypertrophy', *Cell* 153, 4, 828–39. https://doi.org/10.1016/j.cell.2013.04.015

7. Robine, J. M. and Cubaynes, S. (2017). 'Worldwide Demography of Centenarians', *Mechanisms of Ageing and Development* 165(B), 59–67. https://doi.org/10.1016/j.mad.2017.03.004

INDEX

beriberi 126, 131
berries 60–1
beta-amyloid 291
beta carotene 128, 130, 150
bias in publication 76, 196
binge-eating disorder 31
biotin (B7) 131
birch sap water 168
black beans 37
bladder: alcohol effects 177
blood
 clotting 144, 238,
  239–40, 243
 parabiosis 324
 pH level 81
blood glucose
 chromium 135
 effect of consuming
  carbohydrates 22–3
 exercise 204
 fasting 88
 gluten-free diets 54
 glycaemic index (GI)
  23–4
 insulin role 22, 23, 30
 low-protein diet 92
 morning levels 83
blood pressure
 circadian clock 283–4
 exercise 204
 fasting 90
 glucosamine 142
 hormesis 257
 medication 237
 nitrates 65
 potassium 37, 148
 salt 33, 35
 sleep 290
 sunlight 273–4
 veganism 98
blueberries 57, 60–1
blue light 312
BMI 111–12
Body Adiposity Index 112
body fat see fat, body
body temperature 283–4,
 293
bone health 46, 48, 130,
 134, 144–5, 149, 191,
 272–3
brain
 amyloid 291–2
 aspirin 238
 concentration 305
 endorphins 212
 fasting 91
 glymphatic system 291
 ketosis 89
 motivation 307
 reward systems 31

sleep 284–5, 300–1
spindles 299
suprachiasmatic nucleus
 (SCN) 305–6, 309
toxins 94
brassicas 58
Brazil nuts 147
bread 22–3, 25, 34, 52–4
breakfast 82–6
breast milk 44
British Dietetic Association
 94
broccoli 135, 257
burgers 101
bus drivers and conductors
 190
butyrylated-resistant starch
 43

C-reactive protein (CRP)
 90, 254
cabbage 61, 102
cafestol 165
caffeine 85, 102, 164–7, 179,
 284, 295, 296, 300
calcium
 alt-milks 51
 dairy foods 45–6, 48, 51
 haem 44
 osteoporosis 73
 supplements 100, 128,
  134–5
 veganism 100
 vitamin D 138
calcium-channel blockers
 237
caloric restriction 86–7, 92,
 256, 322
calorie content
 fats 11
 food labels 113–14, 117
 sugars 28–9
calorie counting 113–18,
 214
calorie intake
 CICO 113, 213
 individual requirements
  81–2
 population trends 28
 recommended 12, 32
 satiety 43
cancer see also carcinogens
 ageing 251, 317, 319,
  321
 air pollution 261, 263
 alcohol 171
 antioxidants 151–2
 aspirin 239
 butyrylated-resistant
  starch 43

coenzyme Q10 136
coffee 165–6
contraceptive pill 240
dairy foods 43–4, 47,
 48
exercise 190, 192, 200
fasting 86
fish 73
fruit and vegetables 55,
 56, 59
haem 42
hormesis 257–8
HRT 242, 243
IGF-1 47, 90
inflammation 268
iron 144
loneliness 267
meat consumption
 39–40, 41, 42
obesity 109, 112
screening programmes
 231, 244–5
sleep 289, 299
smoking 181
sunlight 271, 274
supplements 128, 130–1,
 133, 138, 146
tea 167
cannabinoids 212
cannabis 184
capsaicin 64, 257
carbohydrates
 added sugars 27–9, 32
 butyrylated-resistant
  starch 43
 changes in advice 20
 composition 20
 digestion 22–3, 27,
  115–16
 evolution of digestion
  96
 fibre 24–5
 FODMAPs 103, 104
 glycaemic index (GI)
  23–4
 low-density lipoproteins
  (LDLs) 16
 official guidance 21, 24,
  27–8, 32
 replacing fats in diet 15,
  20, 78
 starches 20, 22
 veganism 98
 wholegrain 23
 see also sugars
carbon dioxide 161
carbon monoxide 263
carcinogens
 acetaldehyde 177
 acrylamide 25–6

INDEX